D0481730

PLAGUE
TIME

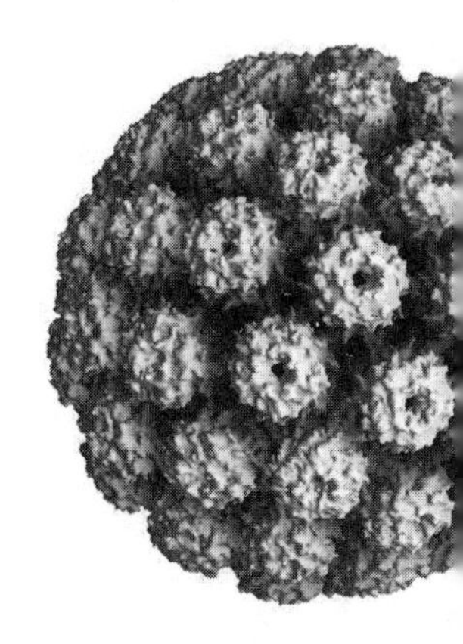

How STEALTH INFECTIONS Cause CANCERS, HEART DISEASE, and Other DEADLY AILMENTS

PAUL W. EWALD

THE FREE PRESS
NEW YORK | LONDON | TORONTO | SYDNEY | SINGAPORE

This book is dedicated to

William D. Hamilton

1936–2000

fP

THE FREE PRESS
A Division of Simon & Schuster, Inc.
1230 Avenue of the Americas
New York, NY 10020

Designed by Lisa Chovnick

Manufactured in the United States of America

10 9 8 7 6 5 4 3 2 1

Library of Congress Cataloging-in-Publication Data

Ewald, Paul W.
Plague time : how stealth infections cause cancers, heart disease, and other deadly ailments / Paul W. Ewald.
p. cm.
Includes bibliographical references and index.
1. Chronic diseases—Etiology. 2. Infection. 3. Communicable diseases.
4. Diseases—Causes and theories of causation. I. Title.

RB156.E93 2000
616.07'1—dc21
00-044249

ISBN 0-684-86900-4

CONTENTS

PART III Beyond the Fear of Infection
Managing Microbes

INTRODUCTION

The Culprits

WHAT ARE WE TO FEAR IN THIS SCIENTIFIC AGE? NUCLEAR WAR? Overpopulation? Drunk drivers? No doubt these are reasonable fears. But threats to our health have a special place in the human psyche. Surely the horror of deadly plagues able to carry away incomprehensible numbers of people has been one of our greatest fears, and certainly is now. But this book is not about the infectious diseases on which the popular media have focused. It is not about the infectious threats to the rich countries from the poor countries of the world—the Ebola and West Nile viruses that captured headlines during the 1990s are in fact minor threats. The balance of evidence indicates that the major infectious plagues are not emerging from an African jungle. They are already here, embedded in every society, in rich and poor countries alike. In fact, they have been here for centuries, even millennia. They are as deadly and painful as the sensationalized plagues, but they have spread more insidiously and imperceptibly—they are slow-motion plagues that are difficult to recognize and difficult to control. The flash-fire outbreaks that make the headlines usually burn out on their own. The serious infectious plagues aren't so easy to escape.

The plague of ancient Athens in 430 BCE, which Thucydides reported, the bubonic plague that transformed the society of fourteenth-century Europe, and the 1918 flu pandemic, which killed more than twice as many people as the weapons of the First World War, were all dreadful scourges of our past. Hardly any sane person thinks that we are suffering from any such calamity now. But we are. We live in a plague time that only a few thoughtful scientists have grasped. Our time of plague will continue for at least the next several decades, even if researchers are unusually clever. The major uncertainty is just how many infectious plagues are smoldering along, consuming the oblivious—those people who think that their bodies are just falling apart from the wear and tear of life. There are certainly several such infectious plagues; there may be many more. A substantial proportion of the pain and suffering that occurred at the dawn of this new century will, I believe, become recognized as the result of long-standing, slow-burning plagues that are infectious but were not generally accepted as being infectious in the year 2000.

This view may seem disturbing and even more frightening than the idea of nasty Ebola-like viruses waiting to catch a plane to London's Heathrow airport, or New York's JFK. But there is good news. Infectious diseases usually have vulnerable underbellies, and recognizing a disease as infectious is one of the most important steps toward discovering its vulnerability. The agents of disease, those enemies within, are often difficult to identify. They are usually invisible and their behavior is deceptive. Although modern medicine can recognize some of these enemies, it has only a few clues about most of the others. The unknown majority are swept under the risk-factor rug, which generates a lot of tabloid journalism—Does drinking some red wine lessen your chance of heart disease? Will holidays in the south of France make you live longer? Should you eat margarine or butter?—while skirting the crux of the problem. We need to understand causes if we are to solve the big health problems.

When major problems are solved, people get a smug sense of superiority over previous generations who failed to recognize what has become an obvious solution. It now seems obvious that surgical instruments should be sterilized before surgery, that water supplies should be filtered and chlorinated to protect against diarrheal diseases, and that condoms can protect against AIDS. Soon it will seem obvious that peptic ulcers are caused by bacterial infection and can be prevented by antibiotic treatment. The task of the present is to recognize that which will be obvious to future generations. This task often requires a mix of skepticism and creativity—skepticism to discern dogma, and creativity to develop alternative explanations that may not fit with the received wisdom. Failing to get this mix right leads to wasting time and resources on the wrong culprits. Bad genes and bad environments have often been falsely accused, or, at least, they have taken more than their share of the blame. Viruses and bacteria are the primary offenders.

• • •

In the mid-1970s I learned a lesson about identifying culprits while I was cutting my scientific teeth, conducting experiments on the territorial behavior of hummingbirds. The study was conducted in one of the dry riparian canyons that were ubiquitous along the coastal counties of southern California before they were transformed into a patchwork of housing developments.

Although there were fewer people in the greater Los Angeles area then, there was no shortage of troublemakers. These were the stomping grounds of the infamous Manson "family." After having a pair of binoculars stolen from my well-worn 1967 Volkswagen beetle, I began keeping a wary eye out for the bad guys. Weeks later, after tuning up the bug, I gave it a quick test drive down the canyon. When a car passed by going the other way, I thought it best to turn around and return just in case. I arrived at my makeshift outdoor work site just in time to stop someone walking away with my tool box: "Sorry, man! I thought that somebody just, like, left it here, man." "Right," I thought, as I vowed to increase my vigilance against thieves.

But my real lesson in vigilance came from a trickier case. The experiments on territorial behavior that I was conducting involved the provisioning of artificial flowers by infusion pumps. The pumps delivered a sucrose solution through a long tube that stretched from a syringe to the artificial flowers from which the hummingbirds would feed. As each pump slowly squeezed the syringe throughout the day, a flower was supplied with the solution. The mechanics sound simple, but Murphy's Law applies with great force to any experiment run outside the walls of a laboratory. After several exasperating problems imposed by weather, friction, evaporation, and the hummingbirds themselves, all was finally working well. Then, midway through an experiment, the vandals struck. When my assistants and I arrived to gather data one morning, the tubing to each flower had been sliced to pieces. Hoping that the act was just an isolated event, we replaced all of the tubing, wrote a threatening sign to ward off the troublemakers, and resumed the equilibration period to allow data collection on the following day. When we arrived in the morning, we found the tubing cut to pieces again. Exasperated and angry, we again repaired the damage, and I wrote a more severe warning about the consequences of interfering with government-supported research, adding that violators would be prosecuted.

Under the heavy influence of adrenaline and testosterone, I returned to the site before dark, thinking that I would wait there and scare off the creeps, hoping that they were little creeps and thinking about what kind of weapons I might be able to improvise if they turned out to be big creeps—the rocks in the streambed perhaps. I was too late. The vandals had already been there, but I must have interrupted them because only a few tubes had been cut. I sat down with tubing in hand next to my impotent warning signs, trying to think what my next escalation should be, keeping an eye and an ear out for anyone who had not yet escaped. Looking at the tubing, I noticed that, strangely, each cut had been made at the same slight angle from perpendicular. Parallel to each cut was a thin scoring of the tubing; it was just one sixteenth of an inch to the side of the cut edge and spanned only half the circumference of the tubing. Having cut the tubing with scissors many times, I realized that scissors did not make that kind of parallel mark. Nor would a knife. Then I guessed, correctly as it turned out, what implement would. I looked around slowly and carefully, and there, near some willows on the opposite side of the stream bank, staring right back at me, was the vandal: a brown towhee. Between its eyes was the destructive implement. Adapted to breaking seeds, that bill could make short work of the strange, long, squishy food with the drippy, sweet reward. The edge of the lower bill severed the tubing as it pressed up against the inside of the upper bill, while the edge of the upper bill left the telltale scoring just next to the cut. I looked back at my warning sign, feeling foolish about threatening a bird with prosecution.

My resolution of the problem had been delayed because I jumped to the wrong conclusion about its cause. Had I thought more broadly about the spectrum of possible causes, I could have resolved the matter much more quickly. On the bright side, it did not take me decades to figure out that my original line of thinking was leading me down the wrong path; and thousands of people did not die as a result of my misguided reasoning.

Unfortunately, the same cannot be said for medicine. Thousands suffered and died because antibiotic treatment of peptic ulcers was generally recognized in 1995 instead of 1955. Thousands more probably suffered and died over a similar period because cervical cancer was treated as bad luck rather than a preventable sexually transmitted disease. It would be gratifying if we could be confident that these oversights were a thing of the past, that lessons had been learned and the health sciences now objectively consider and present the spectrum of feasible explanations for the cause and control of diseases. But the record does not support this optimistic view, not when we assess the last two centuries of medical progress, nor when we assess the last two decades. We have no reason to think that suddenly in the year 2000 the mind-set and the biases have changed.

Cancers, heart attacks, stroke, Alzheimer's disease, and infertility are like the acts of an anonymous vandal. Together they are the primary reason our life expectancy is what it is. Experts say that causation is multifactorial; they consider risk factors but shy away from considering primary causes. Infectious agents are sometimes mentioned, but they are often dismissed without justification. If the true culprit is not suspected, we have little recourse for controlling it. The possible culprits fall into three general categories. One is bad genes. Another comprises harmful noninfectious aspects of the environment such as radiation, noxious chemicals, and dietary imbalances. The third is infectious damage caused by viruses, bacteria, and other parasites—culprits that medicine thought it understood and had under control a quarter century ago. This last category has been grievously underestimated.

Modern medicine is not nearly as far advanced as the textbooks and most physicians would have us believe. It is now more important than ever to identify the weaknesses and mistakes of our medical establishment, not for the purpose of attacking hardworking doctors, but rather to suggest ways to better understand the predicament we are in at the beginning of the twenty-first century and the range of options that we have at our disposal for overcoming the very present danger we face.

PART I

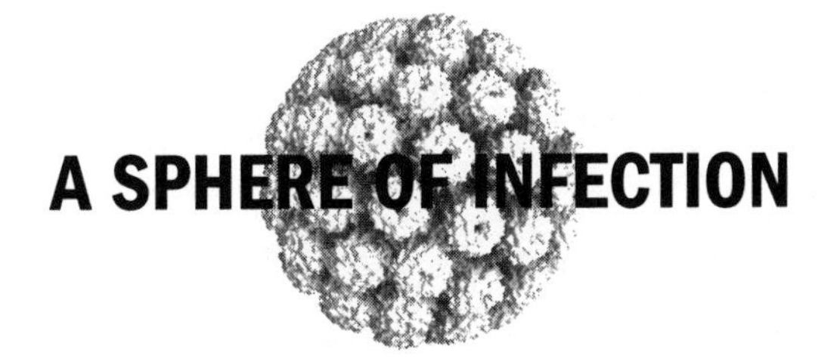

A SPHERE OF INFECTION

The Approach of

Evolutionary Medicine

CHAPTER ONE

The Virulence of Acute Infections

WE ARE THEIR FOOD. THOSE GERMS OF THE PAST THAT BEST converted our bodies into their own propagation are the germs of the present. Those germs of the present that best convert our bodies into their own propagation will be the germs of the future. Why should we care about the prospects of one particular germ over another? Aren't they all just plain bad? The simple answer is no. We can never get rid of them all. Their future is our future. If their future goes one way, we will be relatively healthy; if it goes another, we will be sick or even dead. So the question is, how will they survive? Or rather, how will they evolve?

Surprisingly, neglect of the germ's-eye view of the world is not restricted to the average person; it extends to medicine as a whole for most of its history. Only during the past twenty years have researchers emphasized the importance of looking at a germ's evolutionary scorecard. This scrutiny is suggesting solutions to the most damaging problems of medicine as well as the most irritating. Both categories of problems are important. AIDS, tuberculosis, and malaria are important because they are so damaging; though most of the people reading this book will not suffer from these diseases, they are common enough that our lives are affected indirectly. On the other hand, the common cold is not life-threatening, but it is important because it is such a pervasive nuisance.

Disease from the germ's perspective has been best worked out for the acute infectious diseases. These are the diseases most of us picture when someone mentions infectious diseases—the common cold, strep throat, pneumonia. They typically arise suddenly, within a week or two after the germ has invaded, and are generally controlled by our immune system within a few weeks.

Typically, acute infectious diseases turn quick profits for short-term gain. The pathogens that cause them are corporate raiders, out to get rich quick rather than maintain the health of their targets. If they depend on their host company's well-being, they may have a fairly benign effect. But if the chance to exploit and move on arises, they take it, and the host company suffers and may even be destroyed. Biological parasites take food rather than money, and they spend the food on reproduction rather than material goods. For either kind of parasite—microbial or human—exploitative propensity depends on whether a relatively healthy host is needed for the leap to the next host. When a sick host suffices, the most damaging parasites can prosper.

The germs cannot consciously plan their moves in the way a corporate raider does. But natural selection molds the pathogens so that they act strategically, almost as if they were making plans. The strategic options can be envisioned as a competition that is played out in two contests. The first contest occurs within the host, where the favored competitors are those that most effectively use the host as food for their own reproduction. The second contest is played out in the transmission of pathogens to new hosts; those pathogens that have been successful at growing within hosts are now in competition to reach the remaining uninfected members of the society. A pathogen that never gets transmitted to a new host is doomed; if it is not destroyed by the immune system, its finite future is guaranteed by the inevitable death of the host in which it resides.

These two contests require different talents. The best competitors are those that do well enough at both events to generate the most progeny—thus dominating the next round of the cycle. Natural selection assesses the strengths and weaknesses of competing pathogens much as judges of a decathlon assess the strengths and weaknesses of competitors in different arenas. With natural selection, however, the "points" are copies of genetic instructions. Pathogens earn these points only by propagating through time. A pathogen that takes so much from a host that it compromises its ability to get to the next host may leave fewer descendants than a less gluttonous pathogen. The more frugal competitor may produce fewer progeny within a host but by keeping the host relatively healthy, it may be better transmitted to the next host if, for example, being terribly sick hinders transmission. If a pathogen relies on some well-placed sneeze in an office or a classroom for transport, a person sick in bed would be a disaster. But the converse could also be true. If a pathogen does not rely on a mobile host, then the more gluttonous competitor may have the higher score in the decathlon of evolution. In other words, because the genetically encoded characteristics of a germ that help it win the competition at one stage of the process might hinder it at another, evolutionary biologists consider the trade-offs that are associated with each characteristic. Growing rapidly inside a person typically involves an evolutionary trade-off: the benefit of generating more progeny within a person is weighed against the reduced chances of contacting a susceptible person if the infected person is too sick to move around. Evolutionary biologists are efficiency experts, always assessing benefits relative to costs.

The basic evolutionary principles underlying such analyses have transformed the modern understanding of infectious disease and promise to transform medicine itself because they reveal a fundamental misconception about disease. Throughout the twentieth century, leading authorities in the health sciences believed that coevolution of pathogens with their hosts would inevitably lead to benign coexistence. They arrived at this mistaken conclusion because they did not consider the trade-offs that were a part of the competition. They focused on the long-term survival of particular parasite species as a whole, rather than the success of particular competitors within the species. The mistake is partly attributable to the catchiness of the phrase "survival of the species." With this phrase jingling around the brain like a pop-song refrain, many medical authorities decided that natural selection somehow directly favored the survival of a species. It does not. Rather, natural selection operates through differences in the rate at which certain genetic instructions are passed on relative to different genetic instructions that occur in other individuals of the same species. The eventual survival of the species may be favored or disfavored as a result, but species extinction, if it eventually occurs, is powerless to influence the course of any competition that occurs prior to the extinction. Natural selection obtains its power from the differences in the survival and reproduction of the competitors within a species, which in turn determine differences in the passing on of the genetic instructions that individuals house. That is where one must look if one wishes to understand why infectious diseases are the way they are and what we can do to control them, because that is where the strategies of pathogens are being shaped.

BY WINGS AND BY WATER

Host-parasite associations can evolve to any point along a continuum from extremely lethal to so mutually beneficial that neither participant could survive without the other. Trade-off analyses seek to explain the spots along this continuum to which particular associations will evolve, over millions of years or just a few months. The rate depends on many things: what kind of variation exists among the germs, how long it takes to generate variation that would be useful for the germs, the differences in success among the competing germs, and what the host has and can muster as countermeasures. The outcomes may not be stable.

Of particular importance to the outcomes is the dependence of pathogen transmission on host mobility. If a mode of transmission allows pathogens to reach susceptible hosts even when the infected host is entirely immobilized by illness, then we expect natural selection to favor a ravaging disease. Among diseases transmitted by mosquitoes, for example, even very sick individuals can serve as a source of infection because the mosquitoes take care of the transportation. In fact, sick individuals may be even better as sources of infection because they are less able to swat mosquitoes. As expected from this argument, diseases transmitted by mosquitoes, tsetse flies, lice, and sandflies do tend to be more lethal than diseases that rely on person-to-person transmission. This simple trade-off argument explains why the agents of malaria, yellow fever, and sleeping sickness are so much more incapacitating than agents of respiratory diseases, such as the common cold, which typically cause just sneezes, coughs, and runny noses.

Mosquitoes and other organisms that transport pathogens from person to person are called vectors; the diseases they transport are, logically enough, referred to as vector-borne. As a group, vector-borne diseases are particularly well endowed with killers, including malaria, sleeping sickness, and yellow fever. But even those vector-borne pathogens that are not especially lethal tend to be agonizing and immobilizing. The dengue virus belongs in this category. It is a cousin of the yellow fever and West Nile viruses and is transmitted largely by the same mosquito that transmits yellow fever: *Aedes aegypti.* The dengue virus kills fewer than one of every hundred people it infects, but the low probability of death is little comfort to the dengue patient, as is clear from an account by the tropical-disease expert Alan Spira, describing his own case of dengue, which he acquired in East Africa: "A headache behind the eyes that throbbed and pounded, with a sensation of pressure like a kettle brewing and boiling. A fever, mild at first, but later intense with sweating, came bundled with ferocious muscle aches. These aches were rooted deep in the calves and back, and felt like being punched from the inside-out." Dengue is often called breakbone fever because the pain gives the patient the impression that bones are slowly being broken.

The dengue virus was probably passed to humans from monkeys many centuries ago. Though generally confined to a band within twenty-five degrees latitude from the equator, it has circled the earth within this zone and is found in India, Southeast Asia, Africa, the Caribbean, and Mexico. Using its mosquito transport, it quickly burns through one village and then travels off to another village, town, or even city, wherever the *Aedes* mosquitoes—and hence the viruses—have ready access to their human food. The viruses return to repeat the process when the number of susceptible humans in the group increases sufficiently through new births, new immigrants, or the gradual fading of the immunity conferred by previous infection. Dengue is a terrible experience for the afflicted person, yet the incapacitation serves the dengue virus by making the sufferer a more vulnerable target for mosquitoes.

The dependence of transmission on host mobility also explains why some diarrheal diseases are matters of life or death, whereas others are just an annoyance. People can inadvertently create "cultural vectors," which transport pathogens from immobilized hosts much the way mosquitoes transport malaria and dengue. If water supplies are not adequately protected, the washing of clothes and bedsheets can contaminate the water, thereby infecting hundreds of other people, even if the person who contaminated the materials was entirely immobilized with a case of cholera or dysentery. Waterborne pathogens pay a low transmission price when they exploit a person so intensively that the person is completely immobilized, because they can still be transmitted from immobile hosts; and they gain a big fitness benefit from exploiting infected hosts because contaminated water can contact many more people than an infected person can. Waterborne pathogens are not just limited to an infected person's friends and acquaintances; anyone who drinks contaminated water is a potential victim.

Comparisons of human diarrheal diseases confirm the central prediction of this line of reasoning: the more waterborne the diarrheal bacterium, the more deadly it is. This association explains why cholera, typhoid, and dysentery are so deadly, and the bacteria that infect the intestines of rich countries are generally so mild. This evolutionary perspective also explains why travel to countries without adequate protection of drinking water is so dangerous, even when such countries do not have notorious vector-borne diseases such as malaria, dengue, and yellow fever. The diarrheal pathogen that enters the traveler's body through contaminated food or drink may have had a long evolutionary history of transmission that has not depended on mobile people. The traveler feels this legacy much more intensely than the local residents because the residents have already generated an immunity. These residents typically paid the price of initiation early in life when, as babies or toddlers, their lives depended on the defenses their immune systems could muster. Those youngsters who did not pass this test joined the three million or so children who are buried in poor countries each year, children who died from something as simple and violent as diarrhea. The traveler isn't as likely to die from the infection, largely because travelers usually have access to the simple but life-saving doses of antibiotics and rehydration solution. The traveler lives to ponder this firsthand experience of life as it is lived now in poor countries and as it was lived just a few generations ago in rich countries. In the nineteenth century almost all urban centers had the same nasty diarrheal pathogens. By allowing drinking water to be fecally contaminated, the technology of the time fostered the distribution of deadly agents. The children of rich and poor countries alike experienced the same lottery that is being held in poor countries today. One in ten children typically succumbed to disease in areas with unprotected water. Whenever water supplies were protected, the most dangerous protagonists vanished as predictably as actors at the end of a play.

WHAT PATIENTS GET FROM HOSPITALS

Hospital-acquired infections have their own cultural vectors: doctors, nurses, and other attendants inadvertently transfer pathogens from their patients to their hands and then to other patients, either directly or indirectly through contamination of objects in the hospital. Such attendant-borne transmission is the major route for most serious infections acquired in hospitals, such as staphylococci, streptococci, enterococci, and *Clostridium difficile,* which can cause life-threatening infections of the skin, lungs, and intestinal tract. Attendants usually do not become infected themselves partly because they are less vulnerable than their patients, partly because they will wash their hands before leaving the hospital, and partly because they may have generated some immunity to the hospital organisms. This history is as old as the hospitals themselves.

In the early 1850s the Hungarian physician Ignaz Semmelweis was perplexed by the fact that one out of eight healthy mothers-to-be who were admitted to the University of Vienna hospital to deliver babies were leaving in caskets. The "childbed fever" that killed the mothers shortly after delivery was characterized by sepsis, which in the mid-nineteenth century meant an invasion of the blood by rotten or putrid material. Today it means disease resulting from the presence of microbes or their toxins in the blood, a common consequence of hospital-acquired infections.

Semmelweis's concern turned to horror when he began to understand the reasons for the hospital's alarming statistics. He noticed that the women were dying from the same disease that physicians and medical students had been studying in the morgue in between giving pelvic exams. More important, he noticed that the women who received those pelvic exams were more likely to fall ill than those who did not. He concluded, decades before Pasteur and Koch established the germ theory of disease, that the doctors and medical students were inadvertently killing the mothers-to-be by transmitting some invisible agent of disease during the prenatal exams. After the women were removed to the morgue, the agents that killed them were inadvertently returned to the ward on the hands of medical personnel and then transferred from patient to patient during the pelvic exams. To break this cycle, Semmelweis introduced the practice of having hospital staff wash their hands with a chlorine disinfectant. This treatment was followed by an application of oil to the hands; the oil was intended to serve as a barrier to any organisms remaining on the hands—a mid-nineteenth century version of latex examination gloves.

Within a month, Semmelweis's brilliant intervention cut the mortality in his maternity ward from about one death for every eight admissions to about one death in thirty. He was rewarded in 1854 with the termination of his position. Semmelweis eventually returned to Hungary. Within a few months he was given a position at the University of Pest and was appointed a physician without pay at St. Rochus Hospital in Budapest, where he introduced his life-saving hygienic practices with some resistance, though less than he had faced in Vienna. He died young in 1865 from the same disease he sought to prevent, after accidently cutting himself during a postmortem examination of a woman who had died of childbed fever. In 1969, just over a century after his death, the oldest and most distinguished medical school in Hungary—the one that supported him after the University of Vienna spurned him—was renamed Semmelweis University of Medicine.

While Semmelweis was worrying about expectant mothers in Vienna and Budapest, Joseph Lister was having similar worries about his surgical patients in Glasgow. Half his amputees were dying of sepsis. Over a decade earlier, in 1851, while Lister was finishing his medical studies in London and Semmelweis was starting work at St. Rochus Hospital, the staunch Semmelweis supporter J. F. von Arneth arrived in Edinburgh to present Semmelweis's findings after having been rebuffed by leading Parisian obstetricians. In contrast with the other learned medical societies of Europe, the Medico-Chirurgical Society of Edinburgh responded favorably to the evidence. The 24-year-old Lister moved to Edinburgh a year later. Though Lister missed von Arneth's presentation, the responsive atmosphere in Edinburgh must have fostered the line of thinking that he would soon formulate. The son of a wine merchant, Lister knew of Pasteur's insights about the role of microbes in fermentation, and he made the connection between microbes and hygienic prevention of disease even before Pasteur. Lister became convinced that microbes were the culprits in his postamputation sepsis and that the transmission of the microbes could be blocked by the right hygienic intervention. Looking for a disinfecting agent that was suitable for destroying microbes on living tissue, Lister remembered that sewage had been treated with carbolic acid (now called phenol) in an effort to curb waterborne disease in Carlisle, England, about 90 miles south of Glasgow. On August 12, 1865, the day before Semmelweis's life would be claimed by the bacteria he'd spent his career trying to control, Lister first used carbolic acid successfully, disinfecting a compound fracture in his surgery in Glasgow. He soon developed methods for disinfecting not just the patient but the surgical equipment and the room itself. Deaths in his wards dropped from nearly one out of every two amputees to less than one out of seven during the first three years that his disinfection procedures were in place.

Lister did not suffer the kind of ostracism that stymied Semmelweis, partly because Lister's techniques were more readily demonstrable, but also because he had friends in high places. Pasteur in particular praised Lister's work, much as Lister praised Pasteur's. With support from Pasteur, Lister's techniques were adopted in Paris in 1873, six years after he had published his findings in *The Lancet* and twenty-two years after the Parisian obstetricians had dismissed von Arneth's message. Pasteur finally completed the circle a decade later—nearly a half century after Semmelweis's discovery of attendant-borne transmission—by identifying the streptococcal bacterium that causes childbed fever.

Life-threatening infections are not as commonly acquired from hospitals as they used to be, but they still cause much pain and suffering in both poor and wealthy countries. Depending on the year and how causes of death are categorized, hospital-acquired infections in the United States generally rank somewhere around the tenth leading cause of death. Certainly, many of these deaths occur because patients' defenses are severely compromised by surgical procedures, drugs that suppress the immune system, and other problems, such as severe burns. But just as certainly, attendants could be more conscientious in their hygienic practices. How far have we come? In 1991 a study of a nursery ward for newborns in Chicago showed that nurses followed appropriate hand-washing guidelines about half the time; doctors in the ward followed the guidelines half as frequently as that. Apparently the effectiveness of antibiotics in U.S. hospitals has caused generations of hospital staff to rely on antibiotics rather than on hygiene.

The reliance on antibiotics has generated an arms race with microbes. Most antibiotics are more poisonous to the microbes than they are to us, but clearly, it is in the patient's interest to use the lowest dosage that will eliminate the disease. Within microbe populations, however, there typically exists variation in the vulnerability to antibiotics. Some microbes may have biochemical machinery that binds to the antibiotic, or machinery that destroys the antibiotic, or biochemical pumps that can pump out the antibiotic. If the dosages of antibiotics are not sufficiently high to knock out all of the microbes, the resistant microbes will be left to repopulate the next generation, which will, as a result, be more resistant to the antibiotic than was the previous generation. This increment in antibiotic resistance then forces an increment in dosage. If the dosage is already at a threshold of acceptability, a new antibiotic must be found. This process is an evolutionary arms race: microbes evolve increased resistance, forcing humans to use higher dosages and new antibiotics, which in turn drive increased resistance, and so on.

The problem of antibiotic resistance in hospitals has been widely acknowledged, but the connection between antibiotic resistance and the virulence of hospital-acquired pathogens is often overlooked. Many researchers have had a sense that the antibiotic-resistant pathogens circulating in hospitals were particularly nasty, but when they compared the pathogens isolated within hospitals, the expected difference between the harmfulness of the antibiotic-resistant and antibiotic-sensitive organisms was often not detectable. Antibiotic-resistant *Staphylococcus aureus,* for example, is particularly lethal in hospitals. Much of the damage is attributable to the compromised defenses of patients—burn patients are particularly likely to succumb. This damage is especially evident when antibiotics have little effect. But staph outbreaks also appear to cause more overt disease in nurses than one would expect from studies of the outside community, where although about one third of all people carry *S. aureus,* only a tiny percentage show any sign of serious disease. An evolutionary perspective suggests that comparison of one hospital form with another may be like looking for a lost key in the lamplight—though the hospital environment is the most convenient place to look for representative antibiotic-sensitive strains, it may not be the right place. It may be more informative to compare hospital strains with strains isolated in the community. Such direct comparisons have been sorely lacking.

Although the evolution of virulence in hospitals has been badly understudied, the available information supports the idea that cycling in hospitals makes pathogens more harmful; for example, a review of all hospital outbreaks of the sometimes benign, sometimes deadly diarrheal pathogen *Escherichia coli* showed that the longer the strains had been circulating in the hospital environment, the more lethal the outbreak—strains that had been circulating for just a week rarely caused death, but strains that had circulated for many months killed about one in ten infants.

Though the most frightening hospital strains of pathogens are the highly virulent antibiotic-resistant strains, evolutionary considerations do not suggest that genetic instructions for antibiotic resistance make germs more harmful. Evolutionary considerations suggest instead that transmission in the hospital environment makes germs both more harmful and more resistant to antibiotics. Hospital strains—whether antibiotic-resistant or antibiotic-sensitive—appear to be inherently more harmful than the normal strains that circulate outside hospitals. To remedy this problem, we will need to heed the ghost of Semmelweis: hygienic practices must be improved.

THE FLU IN OUR FUTURE

As late fall turns to winter, perennial concern arises over influenza. Some people are frightened about the flu for good reasons; others have unwarranted fears based on limited knowledge of flu epidemics such as the one that broke out in 1918. The reasonable fears we have stem from the particular vulnerabilities of some groups, such as the elderly, and it is these vulnerable groups that account for the roughly twenty thousand people who die in the United States each year from influenza. That is about half the number of people who die from AIDS, but unlike people at risk for AIDS, those at risk for flu are more evenly distributed across society, so most Americans have had an experience with an elderly relative or acquaintance who has either succumbed to or had a close call with influenza. The danger is similar for most countries in temperate climates.

The unwarranted fears are based on the expectation of an influenza plague as lethal as the 1918 pandemic, during which there were over thirty times as many influenza deaths in the United States as occur in a normal year. About one out of every hundred people on earth died from influenza within twelve months. For the past quarter century, influenza experts have been anxiously searching for new strains that resemble the viruses responsible for the 1918 catastrophe. In particular they have been looking at the form of the two protein molecules that stick out of the surface of the influenza virus. One protein is hemagglutinin, which allows the virus to bind to our cells for entry. The other is neuraminidase, which keeps the flu virus molecules from sticking to each other. These proteins are so closely watched not because the particular form they took in the 1918 viruses made those viruses harmful. Rather it is because the two proteins are the easiest for the immune system to "see." The virus is wrapped in a cell membrane to disguise it from the immune system, and hemagglutinin and neuraminidase stick out from this camouflage the way hands stick out of a shirt. The immune system generates antibodies to the proteins, and scientists can detect the presence of the antibodies. If influenza experts detect a combination of these proteins that is similar to the combination of the 1918 pandemic viruses, which is referred to as H1N1 (for hemagglutinin type 1 and neuraminidase type 1), alarm spreads.

That happened in 1976, causing a cascade of events that flowed out of control. When a large proportion of the population has not been exposed to a particular flu virus, the potential for a global outbreak of influenza is especially great. Flu experts were anxious in 1976 because they were nearing a period when immunity to the H1N1 combination would virtually have disappeared from the population. People who were infected as babies in 1918 and survived were about sixty years old in 1976. So everyone who was less than sixty years old had no immunity to the 1918 virus. The anxiety was magnified by political forces, which pushed through a "swine flu" vaccination program, which caused a paralytic disease called Guillain-Barré syndrome in about five hundred people, about twenty-five of whom died. The benefits of the program are hard to gauge. Probably no more than a few lives were saved from a flu outbreak that apparently burned out on its own.

Concern over a pandemic of an H1N1 virus was justified, but anxiety about a revisitation of a virus with the harmfulness of the 1918 influenza viruses was not. The flu experts of 1976 needed to add the insight of evolutionary medicine to their biochemical perspectives. They reacted too strongly and too quickly to the discovery of H1N1 on the viruses isolated in 1976. It would be like seeing a person with a Hitler-style mustache giving a political speech in 1976, concluding that the person was a threat to world peace, and then initiating attempts to incarcerate him. The similarity in the outward appearance of the 1976 viruses to those of 1918 did not indicate that they would have a similar level of harmfulness. The H1N1 marker had been present on dangerous viruses, but there was no reason to think that it *made* the viruses dangerous—with its high mutation rate, the influenza virus can generate tremendous variation within a matter of weeks while still retaining the same H1N1 marker.

An evolutionary approach to the influenza threat considers conditions that allow transmission from sick individuals. One could hardly imagine conditions as conducive to high virulence as those along the Western Front in World War I. If a person is packed into close quarters with other people, for example, in a trench or in army barracks, and then becomes sick, the others close by would have a good chance of being infected. People who fell sick at the front were typically transported to triage hospitals behind the lines; there the flu viruses in the sick person would have another set of potential victims just a cough away. Within a few hours a sick person would be whisked from the triage hospital to an overcrowded permanent hospital. Some of these hospitals were built to house about two hundred patients but were funneling through two thousand a day during the latter part of 1918; soldiers in these hospitals were packed into rooms and overflowed into hallways. Most were quickly transported to other hospitals—for example, by being stacked in train cars. In such a system an immobilized person could infect hundreds of additional people before his infectious period ended.

But did the first sign of the super-nasty influenza strains arise at the Western Front? In an attempt to understand the origins of the 1918 pandemic, one of the twentieth century's leading experts on infectious disease, Macfarlane Burnet, tried to reconstruct events in a short book with a long title—*Influenza: A Survey of the Last 50 Years in the Light of Modern Work on the Virus of Epidemic Influenza*—which he published with his colleague E. Clark in 1942. They traced the pandemic influenza to army recruits stationed in the United States during the spring and summer of 1918. But the lethality of these infections did not seem to be out of the ordinary. The first cases with the notoriously high lethality were recorded in U.S. troops at the Western Front in the fall of 1918. Both the theory and the evidence therefore implicate the Western Front as the source of the epidemic. This is not the kind of idea that can be tested with controlled experiment. But as long as people do not re-create the opportunities for transmission from immobilized people that were created during 1918, the idea can be evaluated over time by many weaker tests rather than a single strong one, much as the hypothesis of human origins from ancestral apes has been evaluated by many weak tests. If the idea is correct, we will see some things in the future and fail to see other things. We will see flu viruses that have the H1N1 combination but do not have exceptionally high virulence. We will fail to see a recurrence of a pandemic influenza with the kind of lethality that characterized the 1918 pandemic.

In 1990, I first put this prediction in print, emphasizing that we were at that time passing through the period when flu experts were expecting a revisitation of H1N1 viruses. With people born in 1918 being over eighty at the time of this writing, we have passed almost completely into the period of maximum vulnerability to a revisitation of H1N1 viruses. And so far the predictions of evolutionary medicine have proved correct—we have not experienced another epidemic influenza with anything like the virulence of the 1918 infections.

But this evolutionary perspective is still foreign to most influenza researchers. They now regularly incorporate evolutionary reconstructions of the various influenza viruses to determine which viruses are related to which other viruses. But they still confuse the sources of variation—the mutation and recombination of genes—with the process of evolution by natural selection. And they still confuse similarity of hemagglutinin and neuraminidase molecules among different virus strains with similarities in the virulence of these strains. This kind of confusion opens the door to future mistakes mirroring the mistakes that occurred in 1976 and led to the overuse of the swine flu vaccine. By failing to investigate the selective processes that favor increased or decreased virulence of virus strains, experts still run the risk of spending too much time and too many resources in attempts to block a 1918-type pandemic, and too little time on how to deal with the more imminent threats.

LYING IN WAIT

The preceding examples share a common theme. High levels of harmfulness are explained by the opportunities for hitchhiking from immobile individuals. Harmful pathogens can catch rides on mosquitoes, in water, on the hands of hospital attendants, or in the infected person when that person is transported from one cluster of susceptibles to another. There is yet another way in which pathogens can make it from a very sick infected host to an uninfected host: they can rely on the mobility of the not-yet-infected host. They can use a "sit and wait" strategy. If the pathogens are extremely durable in the external environment, they remain where they have been released and wait for the susceptible individuals to come to them. They may also contaminate clothing, blankets, or other possessions that are moved around after an owner's illness or death and lie there in wait for a new owner. Like pathogens transported by mosquitoes or water, durable pathogens can afford to convert more of their host into their own progeny because the immobility of a sick host is less costly to them.

This strategy is especially common among insect pathogens, which can stay viable outside hosts for years or even decades and then make soup out of an insect that happens by. But the strategy also explains why some pathogens of the human respiratory tract are so much more damaging than others. The mildest of human respiratory pathogens, such as the rhinoviruses that cause the common cold, lose their viability after a few hours outside the host. At the other extreme is the smallpox virus, the most lethal of human respiratory tract pathogens. No one knows exactly how long it can last in the external environment. In one study smallpox scabs were stored in an envelope that was left on a shelf in a lab cabinet. By sampling the scabs periodically, the researchers demonstrated viable viruses for thirteen years. They could not continue the study because thirteen years of testing had used up all the viruses in the envelope.

Imagine the number of people who might unwittingly enter a contaminated house or come into contact with contaminated materials over a thirteen-year period. Such durability explains why American Indians in the colonies of New York and Pennsylvania were decimated by smallpox. Lasting for a few days or weeks on the infamous smallpox-laden blankets distributed to American Indians from a colonial outpost would be difficult for most viruses, but not for a virus that could last more than thirteen years on a lab shelf. There were even more morbid consequences of this durability. In 1757, after French forces took over Fort William Henry in northeastern New York, their Indian allies began digging up English graves. They got the scalps they were after, but they also apparently retrieved smallpox viruses that were lying in wait in the corpses of those who had died from the disease.

Pathogens of the human respiratory tract tend to be only as bad as they are durable. The tuberculosis bacterium, for example, can last for weeks to months outside the human body. Before the days of the effective antituberculosis drugs isoniazid and rifampin, tuberculosis was second only to smallpox in its lethal effects on infected individuals. The next most deadly respiratory tract pathogens, those causing diphtheria, whooping cough, and pneumococcal pneumonia, range from being moderately to highly durable in the external environment, maintaining their viability for days to weeks. The mildest pathogens—those that cause little more than some nasal congestion, sneezing, or a cough—typically lose their viability within a few hours of exposure.

A SPHERE OF VIRULENCE

For most of the twentieth century, a poor grasp of natural selection and a tendency to pick and misinterpret specific examples led medical authorities to misunderstand the wide variation in virulence found among human diseases. The poor grasp of natural selection led them to presume that benign coexistence was always the best policy for parasites. The selective picking and misinterpreting of specific examples gave them a sense that their mistaken ideas were supported by the evidence.

Much reference was made, for example, to the high virulence of pathogens that were new to a host. A modern-day example is the Ebola virus, which often causes terrible destruction of tissues but tends not to spread to others in the community. True, the Ebola virus is poorly adapted to us. True, it recently entered humans and would probably be gentler to us if it had had a chance to adapt to us. But these truths do not lead to the conclusion that pathogens will generally evolve toward benignity as they adapt to their host. As I have suggested, a more careful application of evolutionary principles leads to the conclusion that some pathogens will evolve toward benignity and some will not, depending on the trade-offs inherent in the competition among pathogens to use hosts as food and to be transmitted to susceptible hosts. We can expect pathogens that are poorly adapted to us to range across the entire spectrum of virulence, with some being far more mild and others being far more harsh than they would be if they were well adapted. In fact, the overly mild ones should tend to be particularly well represented among poorly adapted pathogens because the immune system has to be able to protect against a tremendous variety of poorly adapted potential pathogens that are raining down on us with every walk in the woods, and every breath in a home. They may be organisms adapted to the cats, dogs, and mice that share our homes, or the fungi and bacteria that grow in the soil, in our bathtubs, or in our faucets, or those that are transmitted in our ventilating systems. If it were not for our immune systems, some of these organisms would turn our bodies to mush. But they have had little opportunity to adapt to us and are easily controlled by our immune systems, which have to be deft at defending against a tremendous array of organisms that would otherwise use us as food. When a few squeak by we notice them. Though such maladapted microbes may have been lucky enough to evade the immunological defenses, their lack of adaptation means that they lack a mechanism for finely regulating their use of the human food they have access to. The result is likely to ensue quickly, and can prove deadly, as with a case of Ebola. If we forget that there are vastly more maladapted microbes that never trouble us than maladapted microbes that do, we may mistakenly conclude that microbes as a rule either die out through overkill or evolve toward mildness, and, hence, that virulence is a state of maladaptation. Too many medical researchers have done just that.

This mistaken interpretation has been erroneously supported by evidence from an Australian campaign to eliminate rabbits. The rabbits were released in 1859 on a ranch in Victoria and quickly increased to such great numbers that they took over grazing land used by agricultural animals and natural wildlife. Australian grazing lands started to bear an uncanny resemblance to arctic tundra during a lemming boom—small mammals scurrying all over the place.

To control the rabbits, microbes were sought. The more lethal the microbes the better. The search led to the myxoma virus, which was not particularly harmful to its native South American rabbits, but killed almost all the Australian rabbits on which it was tested. When the virus was first introduced in the 1950s, it was tremendously effective, killing well over 95 percent of the rabbits it infected. But over the next few decades mortality dropped precipitously as a result of both reduced virulence of the virus and increased resistance in the rabbits. Mortality eventually stabilized at around 20 percent.

Those who believed that pathogens always evolved toward benignity quickly seized on this experience as supportive evidence. The rabbits would disagree. The virus generated by this evolutionary process is about as nasty to the rabbits as the smallpox virus was to us. In fact, the myxoma experience offers no evidence that microbes evolve to benignity. Rather it shows that a microbe chosen as a biological control agent on the basis of extreme lethality coevolved with its new host to a level of virulence that was lower, but nowhere near benignity. Being a mosquito-borne pathogen, the myxoma virus equilibrated at a lethality that was consistent with the range of lethality among vector-borne pathogens of humans, though still on the virulent side of this range.

One of the most interesting points demonstrated by the myxoma experience is that evolutionary changes in virulence can occur very rapidly, over just a few decades. This point, together with some simple evolutionary logic, reveals the incoherence of the traditional argument that pathogens evolve toward benignity. Consider the persistence paradox: How does one explain the great variation in virulence that is present both within and among different kinds of pathogens? Advocates of universal evolution toward benignity were forced to conclude that variations in virulence were just noise in the system. But if harmful pathogens are assumed to be at a disadvantage, they should vanish over relatively short periods of time, just as the most harmful myxoma viruses vanished over only a few decades. If the harmfulness resulted from variation in host susceptibility, the susceptible hosts should similarly vanish over time. Granted this time might be longer, but it should still mount up over generations, just as the reductions in the rabbits' susceptibility mounted up over a small number of rabbit generations. After many thousands of years of exposure to disease agents like smallpox and tuberculosis, one would expect the individuals who are resistant to the disease to make up the entire population.

The way natural selection actually works does not create this persistence paradox. The variation we observe in host susceptibility and pathogen virulence is simply the current stage in a long-running arms race, much like the arms race between antibiotic use and antibiotic resistance. But in this case the race is between the biological weaponry that our bodily defenses impose on pathogens and the pathogens' resistance to them. It is a race between our resistance to pathogens and pathogen resistance to us. Virulence continues because the race continues. As resistant humans become more numerous, pathogens that are not controlled by the human defenses are favored and they begin to prosper. In turn, new resistance mechanisms emerge in hosts.

Variations in the virulence of acute diseases are not just noise in the system. The differences in harmfulness are generally as large as one expects them to be on the basis of fundamental evolutionary principles. The evolutionary framework explains well the variations in virulent acute infectious diseases, and this is a respectable accomplishment because the standard acute infectious diseases make up a large proportion of the diseases that we care about. But the standard acute infectious diseases are not the most complicated of infectious diseases. Acute infectious diseases invade and battle with our defenses. If we are lucky, they are quickly repulsed and we acquire immunity. If we are unlucky, they kill or severely injure us. Whatever the result, the battle and its outcomes tend to be well defined and conspicuous. Our encounters with other microbial adversaries are not so easy to follow. These other microbes may hide out in our bodies for long periods of time and then rise up again at unexpected times, causing new and more horrific reincarnations.

CHAPTER TWO

The Long Fuse of Sexually Transmitted Diseases

SEXUALLY TRANSMITTED PATHOGENS TAKE ON SECRETIVE AND even sinister strategies. The underlying reason is that people have sex with new partners less frequently than they come within sneezing or coughing distance. The sporadic opportunities for sexual transmission put sexually transmitted pathogens in a difficult situation. They have to stay viable within a person until the person has a new sexual partner. They have to persist in the face of an immune system that is superb at recognizing and destroying foreign invaders. Then they must be transmissible to the next sexual partner when the opportunity arises. And that is just to break even. To make a profit, the pathogens must meet these challenges for at least another round of partner change—the more rounds, the better their competitive advantage. Respiratory pathogens typically reproduce to the point of contagiousness and then get wiped out or at least sequestered by the immune system within a week or two. A sexually transmitted pathogen using such a strategy would be cut out of the competition. To evade this fate, sexually transmitted pathogens must employ sneaky tricks.

Sexually transmitted pathogens must act like criminals living in a town that is heavily patrolled by police. Like human criminals, they have developed a variety of strategies for evading recognition, surveillance, and capture. Because the immunological police soon become very familiar with their appearance, most sexually transmitted pathogens keep a low profile.

The bacteria that cause syphilis persist by stripping off many of the external molecules that would make them recognizable to the immune system. They are like criminals who sand off their fingerprints.

HIV impersonates a police officer, wrapping itself in the immune cell's own membrane when it buds off a cell. To stay ahead of the current mug shot, it frequently modifies its telltale features by mutating and recombining its genetic instructions—genetically engineered plastic surgery.

The bacteria that cause gonorrhea are quick-change artists, wearing different external molecules from day to day to avoid being tracked down after they are recognized. They also hide out away from the areas of most active surveillance, causing damage to the reproductive tract but not so much that it generates a sweeping immunological assault.

Herpes simplex viruses keep a low profile by hiding out in the neurons. There they are relatively safe because an immune system that destroys neurons could irreparably cripple the body. When a person is under stress, the viruses in the neurons break their latency and begin producing progeny, which migrate down the neuronal fibers. Presumably stress is an indication that it is a good time to break out of their hiding place—either to get out before something bad happens to the person, or to get out while the immune system may be less able to mount a defense. The tubelike fibers serve them as a kind of subway through which they can leave town without being detected by immune surveillance. The end of the line lies just below the skin. There the viruses disembark and cause the familiar herpes blister, which upon rupturing provides the viruses with their necessary exit. Then they wait for sexual contact to provide access to new bodies and new neurons. If contact occurs and viruses are transferred, the lucky viruses first infect the cells of mucus membranes, perhaps a cell of the cervix, where they multiply. Some are available for immediate transfer; others enter neuron fibers, migrating up them to reach the cells' nuclei, where they can continue their strategy for reaching sexual partners long into the future.

Human papillomaviruses are the organized criminals. They take over cells of the cervix, making them work for the new boss. Instead of obeying the normal rules of self-control that are necessary for each cell to fulfill its assigned role in the body's community, these cells sometimes develop into cancerous growths that spread at the expense of the law-abiding cells. The goal of the papillomavirus is not to create cancer, but to alter allegiances. In the precancerous state, the virus causes the cells to divide at a level that is too high for the good of the body; but that cellular reproduction is good for the virus because it can replicate along with the cells, with little exposure to immune surveillance. The cells are now working for the virus rather than the body, generating a subcommunity with allegiances that conflict with the interests of the larger community. The papillomaviruses are difficult to root out because these cells, being the body's own cells, provide the cover of an upstanding citizen. The viruses may eventually destroy the entire community, as the disregulated growth of an infected cell develops into a cancerous growth, but until the cancer takes its effect, the viruses get resources and turn profits with little exposure. By the time the cancer destroys the body, the papillomaviruses have moved on to infest new communities.

For these pathogens, the disease is not the goal of their activities. Rather, it is the cost of doing business. These outlaw strategies seem tenuous, yet sexually transmitted pathogens are very successful and pernicious in human populations. They are so successful because sex is so seductive. From an evolutionary viewpoint, sexually transmitted diseases are costly to a person, but the cost of no sex is higher. If a person were genetically programmed not to be interested in sex, those genetic instructions would disappear—in about one generation.

LETHAL SEX

Trade-off logic helps explain why some sexually transmitted pathogens are more harmful than others. Competitors that exploit the hosts soon after infection will be able to generate more progeny; they may therefore win out in competition with any other competitors within the host and may be better able to infect any sexually contacted host than they would be if they exploited at a lower level.

They and we pay a price when they exploit intensively, but this price takes on a character that differs from that of the typical acute infectious disease. The tricks that sexually transmitted pathogens employ in their long-term persistence may eventually cause some essential part of the host machinery to fail. The particular part that fails usually depends on the parts of the body that the pathogen tends to inhabit. Because HIV lives inside immune cells, the more aggressive variants of HIV tend to cripple the immune system. Because herpes viruses tend to live in neurons, their aggressive variants tend to cause neurological problems. Because the syphilis bacterium can persist in the face of the immune system, its aggressive variants can invade most organs of the body and cause tissue destruction wherever they go. In the brain they cause mental illness; in the neurons along the spinal cord they cause lack of coordination; in the circulatory system they cause aneurysms.

Once again, this trade-off perspective raises the critical question, What factors favor the more aggressive variants over the benign variants? Evolutionary theory implicates the potential for sexual transmission. In some populations the potential is low because people tend to have fewer sexual partners or use condoms as a means of birth control and for protection from disease. Evolutionary theory suggests that when sexually transmitted pathogens find themselves in such populations, the particularly benign competitors—those that have genetic instructions for curtailed exploitation—are favored.

For the sake of argument, imagine that everyone in a population stayed with the same sexual partner for five years and then changed partners. If a pathogen lost its opportunity for transmission during the fourth year of the infection (because it was controlled by the immune system or the person died), it would be a loser. To break even, a sexually transmitted pathogen must be transmissible for a period extending into the duration of the next sexual partnership. To prosper, the pathogen must be transmissible for periods spanning more than one change in sexual partners. A low potential for sexual transmission should favor pathogens with such restricted activity inside a person that the infection would be transmissible over decades. Their restricted activity would undoubtedly lead to a trickle rather than a flood of pathogens out of the body, but a trickle would be sufficient. Although the trickle would tend to reduce the probability of infection per contact to low levels, under these conditions of high fidelity there would be years of sexual contacts during which transmission could occur. For a sexually transmitted pathogen, success is measured one sexual partner at a time. Until the sexual partnership breaks up, the pathogen cannot do any better than to infect that one partner. A low probability of infection per sexual contact is compensated for by a high number of sexual contacts per partner, and natural selection should favor a benign relationship.

Now imagine the other side of the spectrum of sexual activity. If people were changing sexual partners every week, the benign trickler would lose the competition. The competitors that are programmed to exploit more and reproduce more would be able to capitalize on the few contacts per sexual partner. Each sexual partner might have an increased chance of death or severe damage years down the line, but the loss of a single host is weighed by natural selection against the additional new hosts that could be infected as a result of the more aggressive strategy.

This evolutionary logic leads to an important prediction: sexually transmitted pathogens will evolve increased virulence in populations that have high potentials for sexual transmission. This prediction can be tested by assessing whether it accords with the information on variations in virulence that has been recorded in the medical literature. It does.

HIV, for example, can be subdivided in several ways. The most basic subdivision is into the two main types: HIV-1 and HIV-2. The main group within HIV-1 had its formative period of evolution in central Africa, where the potential for sexual transmission has been very high. The corresponding area for HIV-2 was West Africa, where the potential for sexual transmission has been variable but generally lower. Accordingly, HIV-1 destroys the immune system and brings about AIDS more rapidly than HIV-2. Similar geographic patterns occur within each HIV type. HIV-2 infections tend to be more benign in Senegal, where the potential for sexual transmission is relatively low, than in the Ivory Coast and Guinea-Bissau, where the potential for sexual transmission is higher. The subtypes of HIV-1 that occur in Thailand and in east-central Africa, where the potential for sexual transmission has been high, appear to be particularly nasty.

Similar trends hold for the other major human retrovirus: human T-cell lymphotropic virus (HTLV). Infections in Japan, where the potential for transmission is low, tend to cause lethal cancers and paralysis at a lower rate than they do in the Caribbean, where the potential for sexual transmission is high. In Japan most infections tend to be acquired from mothers, through breast milk. The lethal blood cell cancers, which occur in about one out of every twenty-five infected people, tend to arise late in life; about half occur in people who are over sixty years old. Virtually all these cancers appear to arise from infections that were acquired early in life from mothers. The time between infection and the onset of cancer therefore tends to be about sixty years. In the Caribbean about half the people who develop cancer are less than 45 years old, and most infections are sexually transmitted. The time between infection and cancer is therefore shorter in the Caribbean, perhaps much shorter. The difference isn't just a consequence of living in Japan or the Caribbean. A similar difference is apparent among Japanese Americans and Caribbean Americans, who have presumably acquired their viruses from their respective ancestral lands.

Retroviruses like HIV and HTLV use RNA as their genetic material. Because RNA viruses tend to be more mutation-prone than DNA viruses, one could argue that HIV and HTLV are special cases. It could be argued, for example, that these RNA viruses are able to respond more rapidly than DNA viruses to changes in opportunities for sexual transmission. But similar geographical associations hold for the DNA viruses that have been tested. Genital herpes simplex viruses, for example, have been tested for virulence by inoculating them into inbred lines of mice. As expected from the theory, the herpesviruses from Thailand were more lethal than those from Japan.

The human papillomaviruses (HPVs) are DNA viruses that have both genital forms, which cause cervical cancer, and skin forms, which cause warts. Just a few of the genital forms seem to be responsible for almost all cervical cancer. As expected from evolutionary considerations, women who have more sexual partners are more likely to have the dangerous, cancer-causing genital forms, whereas women who have few sexual partners are more likely to have the mild forms. During the war in the former Yugoslavia, notorious for the use of rape as a weapon, the dangerous genital HPVs spread much more rapidly than the mild genital forms. The more dangerous papillomaviruses appear to be particularly suited for transmission where the potential for sexual transmission is high.

The theory has not been tested within species of bacterial pathogens, but the differences among species show a pattern analogous to that of the sexually transmitted viruses. The most deadly of sexually transmitted bacteria is *Treponema pallidum*, the cause of syphilis. It depends greatly on a high potential for sexual transmission. The agent of gonorrhea, *Neisseria gonorrheae,* and the sexually transmitted chlamydia, *Chlamydia trachomatis,* rarely cause death in infected adults, and their maintenance in populations is not so strongly dependent on a high potential for sexual transmission. This difference in dependence is apparent in the relative success of control measures, a point well illustrated by a program that was conducted from 1990 to 1993 to increase condom use among prostitutes in Fukuoka City, Japan. Condom use increased to four times its original level; *C. trachomatis* infection inched downward by about a quarter, *N. gonorrheae* was down by almost half, and syphilis dropped by nearly 95 percent.

This current state of knowledge about sexually transmitted diseases complements the picture generated by studies of acute infectious diseases. It again supports the idea that damage caused by infectious diseases is probably not just an aberration. On the contrary, the evidence indicates that damaging relationships between us and the microbes that feed on us can be maintained indefinitely when high levels of host exploitation are favored by natural selection. For sexually transmitted diseases, natural selection apparently leads to damaging relationships when the potential for sexual transmission is high. For the diseases that fall more neatly into the acute category, natural selection apparently leads to damaging relationships when transmission can occur from immobilized hosts.

This conclusion has subtle but far-reaching implications for the nature of human disease: infection has more potential than the other possible causes of disease to harm hosts perpetually. Over the long run, even moderately common and harmful genetic diseases will fail to sow the seeds of their own perpetuation, unless they provide some compensating benefit. Without a compensating benefit, harmful genetic instructions can be maintained only at a frequency that is set by their generation through mutation: at equilibrium, the loss in the harmful genetic instructions that results from their harmful effects must match the rate at which the instructions are being generated by mutation. If a genetic instruction is even moderately harmful, then it can be maintained in the population only as a very uncommon instruction. Even if a disease reduced reproductive success by as little as one tenth of 1 percent when averaged over the entire population, the genetic instructions responsible for the disease would dwindle over time. This logic provides deep insight into one of the most important questions we face as a society today: What are the gravest infectious threats for the wealthy countries? The answer is the chronic plagues.

CHAPTER THREE

The Stealth of the Chronic

CHRONOS, FROM GREEK, MEANS "TIME." CHRONIC DISEASES are distinguished from acute diseases because they are drawn out over time. The sexually transmitted diseases (STDs) often have acute phases, but almost all of them have a chronic phase. Their acute phase often involves a lesion, pain, and inflammation near the site of entry a few days or a few weeks after the onset of infection. But STDs cause their worst damage at more distant sites as a result of their chronic presence. This continuum between acute and chronic disease among STDs reveals a surprising inconsistency in generalizations about disease causation. STDs, and a few other diseases, such as tuberculosis, that demonstrate this continuum, have led to the recognition that infectious diseases can be chronic. But when it comes to chronic diseases that do not have a distinct acute phase, infectious causation is often either dismissed or not even considered. Peptic ulcers, for example, do not have a distinct acute phase, and the infectious causation of ulcers was dismissed for a century in spite of supportive evidence.

A shocking realization arrives when we notice the general trend of which peptic ulcers are merely a specific example. All the diseases that were accepted as infectious during the last quarter of the nineteenth century were either entirely or largely acute—the diseases were obvious because sufferers had obvious symptoms just after they were infected—but all the human diseases that have been accepted as infectious during the past quarter century have been entirely or largely chronic. They are the stealth infections.

Knowing the reasons for this turnaround provides a sense of what the future holds in store, and how we can make that future more healthy. But to fully understand these reasons we need to be part psychologists and part medical historians.

SEEING THE INVISIBLE IN THE NINETEENTH CENTURY

Infectious causation of some acute diseases was recognized early in the 1800s as a result of conspicuous chains of transmission. The infectious nature of smallpox, measles, and chicken pox was recognized by medical experts and the general public decades before the microscope led Koch, Pasteur, and other early microbe hunters to the first cause-and-effect linkage of bacteria with disease during the 1870s and 1880s. Before the identification of microbes, the concept of infection was less tangible, and the distinction between infection (the invasion and growth of a disease-causing entity in the body) and contagion (the spread from one body to another) was often blurred. But infection was invoked to refer to diseases caused by something that grew inside people and could be transmitted to others to continue the process.

No special training was needed to witness the chains of transmission of many respiratory tract diseases, only a moderately observant eye. Parents noticing chicken pox in their child's playmate would see chicken pox in their own child within a couple of weeks, and then chicken pox in another playmate within a couple of weeks after that. They wouldn't have to be a Pasteur to realize that something was growing inside the bodies of their kids and being transferred from child to child. Though the viruses that caused these diseases were not seen with a microscope until the middle of the twentieth century, a half century after Pasteur, the medical texts before Pasteur's time already had the causation and transmission right for these diseases. These texts had it wrong, however, for every other category of infectious disease.

At the end of the nineteenth century, after the microbe hunters of Pasteur's time had observed some of the bacteria that cause diarrheal diseases, the writers of medical texts were still arguing about whether the bacteria caused the diarrheal diseases or were just innocent bystanders. The confusion can be seen in the standard American medical text of the 1880s, *A Dictionary of Medicine.* Three decades after the transmission of cholera had been neatly demonstrated by the London physician John Snow, and a few years after the bacterial agent of cholera had been identified by Robert Koch, the *Dictionary* considered "Asiatic cholera" to be infectious, but transmission rather than infection was emphasized. "Simple cholera" was considered nontransmissible. This distinction probably arose because Snow's work left little doubt that epidemic cholera—which could be traced to south Asia and was therefore termed Asiatic cholera—was waterborne. Snow deduced that the cause of the disease must be some organism that could multiply inside people, be shed in the feces, and be transmitted in fecally contaminated water and food. No one had done similar work on "simple cholera," which probably included what are now recognized as several different diseases characterized by watery diarrhea. So those who favored noninfectious explanations for diarrheal disease could justify relinquishing only the epidemic cholera. The *Dictionary* noted the presence of the typhoid bacterium in typhoid patients but attributed typhoid fever to transmissible "typhoid poison" rather than transmission of the organisms themselves.

Why were acute diarrheal diseases recognized so much later than the acute diseases of the respiratory tract? A simple exercise can reveal part of the answer. When you go to work, school, or wherever you mingle, study the people next to you. Look and listen for signs of respiratory tract infection. If someone is infected with, say, one of the rhinoviruses that cause the common cold, you will have a good chance of detecting it. You may see or hear them sneeze. If they talk, you may hear the telltale muffling of nasal congestion. Then look and listen for signs of diarrheal disease. It's not so easy. Other avenues of information are also restricted. Say you see a not-so-close acquaintance who looks a little peaked and ask, "Feeling all right?" You might get the response, "Oh, I've got a cold." But how many times have you heard "Oh, I've got diarrhea"? People prefer not to broadcast information about such matters. It tends to be reserved to very close friends and family. It is not surprising that infectious causation of diarrheal diseases, being reinforced by less evidence from common experience, was generally accepted later than infectious causation of acute respiratory tract diseases.

Infectious causation of diarrhea was probably also more difficult to recognize than infectious causation of respiratory tract diseases like chicken pox because many people who are infected with diarrheal pathogens show no symptoms at all. Diarrheal diseases also can be transmitted over long distances—for example, by fecally contaminated water. So if one is assessing whether diarrheal diseases are a result of something growing in a person and being transmitted to other people, one will often find seemingly contradictory information. If one chooses as a study subject someone who acquired an infection from an asymptomatic person or through the water supply, no direct contact with diarrhea may be apparent, no matter how careful and thorough the study.

Acceptance of infectious causation of vector-borne diseases was delayed for similar reasons. If a disease is transmitted by a mosquito, an observer might track down every contact of a sick person without turning up another person who has the disease. That is just what happened in a study of yellow fever in Barcelona. In 1822 the French government was still trying to learn the lessons of the Haitian revolution that had occurred two decades before. The Haitians were successful in large part because yellow fever killed thousands of French soldiers while leaving

Haitians relatively unscathed. Wanting to figure out the problem before their next clash with yellow fever, the French government sent Nicholas Chervin to Barcelona to determine whether yellow fever was contagious. Chervin's careful studies documented that many people who came down with yellow fever had no contact with anyone else who had yellow fever. He concluded, logically enough, that yellow fever was not transmissible between people. He did not consider other possibilities for long-distance transmission. If one does not entertain the full range of possibilities, one may miss the right answer. Infectious causation of vector-borne diseases remains obscure if one does not consider the possibility of vector-borne transmission. The Cuban epidemiologist Carlos Finlay did so six decades later, correctly implicating *Aedes* mosquitoes as the vector for yellow fever. Still, the idea of infectious causation of yellow fever was often rejected out of hand. In 1898, for example, an official bulletin of the U.S. Marine Hospital Service stated that "one has not to contend with an organism or germ which may be taken into the body with food or drinks but with an almost inexplicable poison so insidious in its approach and entrance that no trace is left behind." When one isn't on the right track, problems do seem inexplicable. Acceptance of yellow fever's infectious cause finally occurred a few years later, after Walter Reed's commission published the results of its experimental transmission of yellow fever to humans.

"Oh, I've got diarrhea" would be an unusual response to a casual "How are you?" but "Oh, I've got a venereal disease" may never have been given. That kind of information is withheld even from close friends and family—sometimes especially from close friends and family. One might try looking and listening for evidence of sexually transmitted infections among casual acquaintances, but as with diarrheal diseases, telltale signs are lacking. Sexual transmission also introduces a novel source of crypticity. Because few people have sex in public and most do not go around broadcasting their sexual activities, knowledge about who has had potentially transmissible contact with whom is lacking.

Not surprisingly, the infectious causation of sexually transmitted diseases was also recognized later than that of respiratory diseases. The *Dictionary of Medicine* attributed gonorrhea to causes such as too much physical disturbance of the genitalia by too much sex or masturbation. The presence of *Neisseria gonorrheae*, which had been discovered five years earlier, was mentioned but was interpreted as possibly being that of an innocent bystander. Syphilis was referred to as "a specific contagious, noninfectious disease; communicable by contact of the poison with a breach of surface, or by hereditary transmission." The delay between pathogen identification and acceptance of infectious causation was even longer for *Trichomonas vaginalis*, which was isolated and described in 1836 but was not generally accepted as a cause of vaginitis for a century.

GRASPING THE CHRONIC IN THE TWENTIETH CENTURY

Infectious diseases are diverse. They are diverse in transmission modes, diverse in their use of host tissues, and diverse in the harm they cause. Medicine understands the acute infectious diseases fairly well because the chains of infectious transmission range from being very conspicuous to pretty conspicuous. A few, such as smallpox and malaria, cause terrible problems for people. But the vast majority rarely kill, and most are so mild that the fitness costs they impose on humans would not be sufficiently high to implicate infection.

No one grasps yet the distribution of virulence among chronic infectious diseases because the health sciences are now still in the midst—or perhaps at the beginning—of discovering the scope of infectious causation of chronic diseases. The germ theory has been widely accepted since about 1880. During the first century of this period, almost all the recognized chronic infectious diseases had a distinct acute phase. Diseases like tuberculosis had chronic phases that were easy to link to infectious causes because the development of chronic disease involved a slow and observable transition from the acute phase to the chronic. The chronic phases of diseases like syphilis were a bit more difficult to recognize as being caused by infection, because of gaps between the acute and chronic phases and because the different phases had fundamentally different symptoms. The acute phase was characterized by lesions on the genitalia, the first chronic phase by a generalized rash, and the late chronic phases by such a variety of disease states, including heart disease, insanity, and paralysis, that syphilis became known as the great imitator. A chronic disease such as shingles was still more difficult to link to its infectious cause because it surfaced after a long disease-free period had elapsed after the acute phase, which we call chicken pox; moreover, shingles does not occur in everyone who has had chicken pox—some people die before it would have occurred, and the immune systems of other people have apparently so thoroughly controlled the virus that it cannot resurge in the form of shingles.

During the first three decades of the germ theory, from about 1880 to 1910, the scope of acute infectious diseases was quickly resolved, and hypotheses for infectious causation of chronic diseases were advanced. Chronic diseases that were the most easily linked to their acute beginnings were broadly accepted during this period as different manifestations of specific infectious processes.

During the first half of the twentieth century, medical researchers confirmed that infections caused various chronic diseases that appeared as delayed consequences of acute diseases with entirely different symptoms. In 1909 the Hungarian pediatrician J. von Bókay provided evidence that convinced many of his colleagues that the hypothesis he first published in 1892 was true: shingles is a delayed manifestation of chicken pox. After about a half century of observation, experimentation, and debate about rheumatic fever, it was finally accepted during the 1940s as a delayed, often chronic manifestation of previous infection with *Streptococcus pyogenes,* the primary agent of strep throat.

By the middle of the twentieth century medical science was poised to move into an even more cryptic realm of the spectrum of infectious disease—those chronic diseases that were caused by infections that did not generate obvious acute phases. Then medical science dropped the ball. A few people noticed it, and kicked it around a bit, but for the most part it was left alone. This period strikingly parallels the three decades before the first flowering of the germ theory in the nineteenth century, when Jacob Henle's call for investigating infectious causation of acute diseases was similarly dismissed without any evidence to justify the dismissal. It is not entirely clear why medicine dropped the ball between 1950 and 1980. In the 1940s the hypothesis for the infectious causation of peptic ulcers, cardiovascular disease, and cancer was still being considered. In some cases people were even being cured with antibiotics. A combination of developments in science and medicine were misinterpreted and misapplied as leaders failed to guard against the biases of human thought.

Ironically, one of the reasons for this slowdown was the same reason for the tremendous success in identifying infectious causation during the preceding three quarters of a century: the adherence to Koch's postulates. In a presentation to the Tenth International Medical Congress held in Berlin in 1880, Robert Koch set out powerful guidelines for identifying infectious causation. These guidelines, which have come to be known as Koch's postulates, were simple rules for maintaining a high standard of evidence in an area of research that was burgeoning at the end of the nineteenth century. Koch advised researchers to (i) demonstrate the putative pathogen in each patient with the disease, (ii) recover and grow the pathogen in pure culture, (iii) produce the disease in humans or laboratory animals with the cultured microbe, and (iv) recover the pathogen microbe from the diseased animals. The historical record shows that when these guidelines were met, the diseases invariably turned out to be caused by infection. This success then led many experts to insist that infectious causation be accepted *only* when these guidelines had been met—and that is where the logical error crept in. The goal of the experts was laudable; they were trying to ensure that medical research would be rigorous. But they presumed that if a batch of evidence is sufficient for the acceptance of the validity of an argument in one situation, then other kinds of evidence are insufficient for the acceptance of the validity of the argument in other cases. Koch did not make this error; he cautioned that researchers should not use these guidelines as the only basis for ascribing infectious causation. Unfortunately, it is easier to follow guidelines than to think critically on a case-by-case basis, and the experts of the mid-twentieth century set in stone Koch's postulates as the standard for ascribing infectious causation. The consequence was that even when important evidence of infectious causation was available, it tended to be dismissed if Koch's postulates were not satisfied. This dismissal had major consequences for understanding the scope of infectious causation because for some infectious diseases Koch's postulates can be virtually impossible to fulfill.

Cancers are a paramount example. Consider adult T-cell leukemia, the lethal disease that results from a cancerous growth of white blood cells. This cancer has been especially well studied in Japan, where people who die from it are infected as babies from their mothers' milk. Though infected during the first year of life, they first develop leukemia many decades later—about half the people who eventually develop the cancer do so after their sixtieth birthday. Only about one out of every twenty-five infected people develops the cancer. Imagine trying to apply Koch's postulates to evaluate whether suspected viruses cause this cancer. Human subjects cannot be used for ethical reasons. Even if they could, who would conduct a study that might take sixty years to complete, and who would fund it? An agent of such a disease might cause the disease only in humans, precluding the use of laboratory animals. If an agent does cause such a disease in laboratory animals, the disease would have to be different if only because lab animals do not live sixty years. If the disease is different—for example, if it develops more rapidly—one can always argue the laboratory model is not generating the same disease and is therefore not trustworthy. Just this kind of argument was used by cancer researchers during the early decades of the twentieth century to dismiss the relevance of Rous sarcoma virus, which was shown to be an infectious cause of muscle cancer in chickens in 1909. It is being used now to dismiss the relevance of mammal models for breast cancer. The body of evidence in lab animals supports the idea that mammary tissue cancer is caused by viruses. Although the evidence from humans is consistent with this idea, viruses are still broadly dismissed as a primary cause of breast cancer. Genetic causation, however, is presumed, even though current evidence suggests that genes are responsible for at most only about 10 to 20 percent of all breast cancers. Moreover, the genes associated with chronic diseases may turn out to be genes that make an individual susceptible to the infectious cause of the disease. Medical people who dismissed infectious causation of breast cancer during the 1990s may appear as myopic to historians in the year 2020 as those researchers in the 1970s who dismissed even the possibility that anything more than a trace of human cancers could be caused by infection.

And that brings us back to the purpose of this historical jaunt. We need to recognize a long-standing trend that continues to the present. The cases of infectious causation that have been accepted at any particular time during the past two centuries have mechanisms of infectious causation that tend to be more cryptic than those of diseases that had previously been recognized. Infection has been found again and again to be the cause of chronic diseases previously thought to have been caused by defective genes or noninfectious environmental agents such as radiation and chemical pollutants.

Few new examples of infectious causation were accepted from about 1950 to 1980. One of the reasons for this slowdown was the equating of acute diseases with infectious diseases. This error was explicitly incorporated into the policy and goals of this period. In 1967, when the U.S. surgeon general William H. Stewart made his infamous statement about closing the book on infectious disease, he was actually advocating a shift of attention from infectious diseases to chronic diseases. Of course, if chronic diseases are caused by infection, the proposed shift away from infectious diseases makes no sense. Funding was switched to chronic diseases under the hidden assumption that the viable hypotheses for causation of chronic diseases excluded hypotheses suggesting infection. The progress made on preventing diseases slowed almost to a standstill—in spite of vastly greater financial investment. The U.S. National Institutes of Health, for example, spent twice the amount of inflation-adjusted dollars on research in 1990 that it spent in 1970, and will spend about twice as much in 2000 as it spent in 1990.

Nixon's War on Cancer during the early 1970s was an exception to the rule that funding of chronic diseases has neglected infectious causation. The War on Cancer was roundly criticized during the late 1970s and 1980s, and by many in the 1990s, as a failure that occurred because medical science knew too little about the basic biology of cancer to make good use of the money. Now, with a quarter century of hindsight, we can see that this criticism was at least partly false. Those few years of generous funding allowed research dollars to flow even to those who were investigating infectious causes of cancer.

During the 1960s and 1970s cancer researchers were divided into camps that took an either-or attitude. Cancer was attributed to noninfectious agents and human genes or to infectious agents, but rarely to a combination of all these factors. There was no evidence then and there is none now to justify this divided approach. Yet it persists largely because people confuse evidence favoring one hypothesis with evidence against an alternative but consistent hypothesis. A hypothesis of infectious causation cannot reasonably exclude noninfectious influences—all infectious diseases are influenced to some extent by genetic and noninfectious environmental factors. Similarly, evidence for genetic causation does not exclude a role for infectious causation. Still, the discovery of oncogenes (genes that are directly responsible for causing cancer) and their generation from other genes through mutation led many to make this error in logic, to reject hypotheses of infectious causation without any evidence to justify the rejection. Genetic and environmental risk factor research became the fashionable new frontier for research during the 1980s and 1990s.

The various factions are still fighting for funds and recognition, but now we are in a position to look back at the track records of these various camps. This kind of research is supported at a very high level relative to basic scientific research largely because it promises to solve health problems. It is therefore appropriate to assess these track records in the context of improvements in health. The genetic camp made important contributions to basic biology. They are still making promises, however, about how their approaches will improve human health, holding out hopes, for example, for genetic manipulation. These hopes may be fulfilled, but no practical solutions to cancer have yet been generated by genetic manipulation.

In contrast with this lack of practical success, those who were studying infectious causation of cancer have made tangible improvements in human health over the last quarter century, particularly by demonstrating the value of reducing the transmission of infectious agents. Any woman who so chooses can now reduce her risk of cervical cancer by using barrier contraceptives and by having fewer sexual partners, because these activities reduce the chances of becoming infected with the papillomaviruses that cause cervical cancer. Anyone who receives a blood transfusion today has a reduced risk of liver cancer because the blood supply is now protected against hepatitis B and C viruses, which were shown to cause liver cancer during the last quarter of the twentieth century. Anyone who wants to reduce the risk of stomach cancer can do so by eliminating *Helicobacter pylori* through antibiotic treatment. The list of tangible successes goes on, and appears to be expanding to include several cancers that appear to be on the verge of being ascribed to infectious causation, such as breast cancer and colon cancer.

The return from studies of noninfectious causes of cancer is more tricky to evaluate. Reduction in smoking has been by far the greatest success story, but the link between cancer and smoking was known during the first half of the century, very soon after cigarettes were introduced. The major success during the last half of the twentieth century involved socio-politics more than scientific discovery—how to get people to quit smoking and how to counter the powerful vested interests that encouraged smoking. Unlike the situation with infectious diseases, there are no noninfectious environmental factors that seem to be on the verge of explaining much of the cancer that still lacks a suitable causal explanation.

These unexplained cancers amount to about three quarters of all cancer. Infectious causation now accounts for about 15 to 20 percent of human cancers, and suggestive evidence implicates infectious causes for most of the remainder. Less than 5 percent of all cancers are known to be caused without any assistance from infectious organisms.

THE EVOLUTIONARY POWER OF THE CHRONIC

There is a crucial unrecognized problem with genetic causation of harmful diseases, whether they be cancers or other chronic diseases. How do the harmful genes maintain their representation over time? The harm caused by a gene reduces its representation in the next generation in proportion to the negative effects of the gene on the survival or reproduction of the people who carry it. Over generations, a harmful gene will become so rare that it will be present in numbers no larger than what one would expect as a result of ordinary genetic mutation. Because such mutations occur at a low frequency, genetic diseases cannot cause much harm in the population as a whole. Recall that even if the negative effect caused by a harmful gene amounted to a reproductive loss of only one tenth of a percent when averaged over the population, the gene would be too damaging to be maintained by mutation alone. Chronic diseases, if they are common and damaging, must be powerful eliminators of any genetic instruction that may cause them. Only if a genetic instruction provides some compensating benefit can the disease it causes persist as a common ailment. As a rule, chronic genetic diseases can persist as common if the genetic instructions protect against parasites. Genes coding for the chronic disease sickle cell anemia, for example, are maintained because they protect against malaria.

The harm caused by infectious agents is not constrained in this way. Infectious organisms can inflict much greater competitive costs on their hosts, and they can do this indefinitely because their success comes at our expense. Remember, we are their food. The conflict of interest between pathogens and us is therefore like the conflict of interest between a lion and a zebra, or between caterpillars and an oak tree. Consumers can impose great harm on the organisms they consume because the conflict of interest between them pits each consumer in an evolutionary arms race with the species it consumes. Chronic diseases are powerful evolutionary forces here, too, but their power to favor evolutionary increases in resistance to the pathogens is matched by the power of the pathogens to change their strategy for exploiting the host. The damaging diseases can therefore persist indefinitely. If we see chronic diseases that have commonly been causing damage for a long time, the best bet is that they have infectious causes.

Like many great ideas in biology, the idea implicating infectious causation in chronic diseases, though simple, has far-reaching implications. It is so simple and so significant that one would think it would have been recognized by many and would be the starting point for any discussion of the causes of disease. Not yet. As an evolutionary biologist, I cannot claim credit for it either. It was suggested to me by my creative and eccentric colleague, Gregory Cochran, a physicist who spends most of his time resolving intriguing conceptual problems in biology and medicine.

No one can yet say how broad the scope of infectious causation of chronic disease is. Some simple steps of logic, combined with principles of evolution and genetics, lead to the conclusion that most of the highly damaging chronic diseases, including most cancers, are caused by infections. Predictions about infectious causation of mild chronic diseases are much more speculative. The chronic diseases mentioned above are probably just the tip of the iceberg. It is the damage they impose that indicates they are probably caused by infection. But this conclusion does not exclude the possibility that mild chronic conditions, too slight to earn the rank of disease, are also caused by infection. On the contrary, the iceberg metaphor may well apply even though the negative effects on human fitness of these chronic conditions are so slight that they could in theory be maintained by mutation. Some of the uncertainty can be resolved by looking at identical twins. If pairs of twins tend not to share a given chronic condition, we know that we have to look beyond genes to understand the problem. That leaves us with only two alternatives: infection or some noninfectious environmental influence.

The track record on mild chronic diseases also lends credibility to a much broader scope of infectious causation. Dandruff, acne, warts, halitosis, athlete's foot, and gingivitis are mild chronic conditions now known to be caused by infection. Each was widely interpreted as the natural wear and tear of life or the natural imperfection of humans long after the germ theory of disease had been established.

Naturally failing immune systems and naturally overactive immune systems have been blamed as noninfectious causes of chronic disease. But does it make sense that evolution would have generated such a complex system of checks and balances if these complexities were only to generate the system's downfall? Imagine making an extravagant security guard system that self-destructs without provocation. We would not expect people to select the ineffective complex system over a more effective simple system. If an extravagant immunological security system self-destructed, then natural selection similarly would favor the simpler system. The more extravagant system wouldn't be reproduced. We can expect that natural selection will not favor added complexity to immunological security systems when the increased complexity destroys the body without provocation.

The immune system may, however, be driven to self-destructive tendencies through the ever-changing strategies of the burglars it is attempting to deter. *Streptococcus pyogenes,* for example, may damage the heart because one of the germ's proteins has evolved camouflage that makes it look so much like a protein on the human heart that the immune response against the bacteria attacks the heart. In the absence of such provocation, though, the immune system is the quintessence of efficiency and flexibility.

CHAPTER FOUR

The Magnificent Defense

MENTION THE "NATURAL ENVIRONMENT" AND MOST PEOPLE think of the plants and animals that exist in places not yet overrun with humans. Our minds perceive the natural environment and the human environment as very distinct—an old-growth redwood forest is put neatly in one category and a New York City subway in the other. The natural environment in which we evolved consisted of savannas, meadows, and forests. Now we live in enclosures of steel, glass, plastic, and plaster. Most of the animals that share these unnatural environments with us—our dogs and cats—are domesticated; that is, they are different from their wild ancestors, genetically altered through human influences on their breeding. In this sense they are all unnatural. So too are virtually all the plants and animals we eat, whether or not their packaging dubs them "natural." Many of us spend most of our waking hours looking through glass plates at the changing patterns of light generated by electrical current. It is easy to come to the conclusion that we have left virtually all of the natural world behind in our movement toward an ever more technologically sophisticated civilization.

That conclusion is wrong. We accept it because our perceptions of nature are profoundly biased by our size. Being large animals, we easily recognize all the organisms at the large end of the spectrum of life. We are hopelessly inept at recognizing in our daily experience the far greater number of organisms so small they cannot be seen with the naked eye. These microscopic organisms are parts of nature too. And when we moved from caves made of rock to caves made of wood, metal, plaster, and glass, we did not exclude them from our daily existence in the way that we excluded lions, wolves, eagles, and frogs. Some microbes have been added to our immediate environment as a result of our closer association with domestic animals over the past ten thousand years. Only a tiny minority have been excluded—those that were sufficiently harmful to us to make us care about them, and were sufficiently vulnerable to be knocked out by antibiotics or vaccines. The rest have come along for the ride, often changing, to be sure, but changing in ways that are not fundamentally different from the ways they have been changing throughout our quarter-million-year tenure as a species on earth.

Our intimate natural environment has always included an abundance of invisible organisms. Now we are familiar with the tiny minority that cause us obvious problems, and we are slowly but increasingly becoming aware of the rest—as microscopes make visible the invisible, and molecular techniques reveal their chemical footprints. Still, we continue to ignore microbial wildlife because most of it does not cause obvious problems for most of us.

This comfortable state of affairs is made possible by henchmen that mercilessly identify, tag, poison, blast, and eat the microbes that trespass on our biological turf. The number of these henchmen within each of us is greater than the number of people on earth, and they are organized into one of the most remarkable inventions that has graced the globe: the immune system. After studying the immune system for about a century, we still do not understand just how remarkable it is. Our minds, at least at present, have proved too feeble. Like the brain, the immune system stores information, communicates, and makes decisions. It also patrols, enforces, and attacks with an efficiency and ruthlessness that make the Gestapo look quaint.

The discovery of antibiotics is one of the great achievements of medicine. But these medications are like children's toys compared with the extraordinary complexity of the immune system's miniature enemy-detection sensors, communication systems, and teams of specialists. These specialists are more diverse and more flexible than the members of any police force. The best specialists are selected for the job. The numbers of these specialists are increased as necessary according to communiqués from other specialists. New kinds of specialists are created when the old specialists do not quite have the right abilities. When the mission is accomplished, the force of unnecessary specialists is reduced to just a few of the best, who are poised to go through the whole process again more quickly if the same kind of adversary shows up.

These teams of specialists include individuals that make specific tags (antibodies) that are put on microbes so other members of the army (macrophages and other phagocytic cells) can recognize, surround, and capture each invader. The invaders are then disposed of with chemical weapons such as peroxide. Some specialists, such as the macrophages, take body parts of the engulfed pathogens and mount them on stalklike structures on their surface much like the victors in human conflicts mounted the heads of their victims on pikes. This "antigen presentation" sends a powerful message. Other cells, called helper T cells, contact the presented body part to see whether it fits their own recognition machinery. If the fit is tight, the helper T cell then reproduces itself prolifically; the progeny scout out other cells that can also recognize the specific enemy but have different talents at their disposal. When the helper T cell finds the same microbial body part on another type of immune cell, the helper cell says in chemical language, "Yes, you have found the enemy. Now use your particular expertise in the control effort." Some of these other cells are demolition experts, called cytotoxic T cells, which then reproduce and move through the body to find infected cells. Infected cells will mount body parts of the pathogens on their surfaces, much as the macrophages do, but this mounting indicates their infected state—they are marking themselves for destruction. When a cytotoxic T cell with the right fit binds to the mounted body part, it blows up the infected cell.

One of the other major targets of the helper T cells are the B cells, which transform themselves into antibody-producing machines after contact with a T cell that carries news of a pathogen. The news is conveyed by an actual part of the destroyed pathogen that locks onto an antibody anchored on the B cell surface. The antibodies produced by these transformed B cells are the same molecule the B cells used to receive the news, but now they are released from the cell into the blood, lymph, and sometimes into the tears and saliva. When the antibodies encounter the pathogen part, they attach to it. When the pathogen part is still attached to an intact pathogen, the pathogen becomes coated with antibodies; it thus becomes recognizable by phagocytic cells like macrophages, which engulf and digest the mess.

Even with all the complex communication, mobilization, and destructive power, pathogens would still often overwhelm the immune system if the B cells of our fish ancestors had not evolved a clever trick hundreds of millions of years ago. The pathogens' inherent advantage derives from their short generation time. The mutations and genetic rearrangements of infectious agents, coupled with their short generation time, provide them with a tremendous potential for staying well ahead of humans in the race between offenses and defenses and counterdefenses and so on. To deal with this disadvantage, the cells that generate antibodies play the pathogens' game. They reproduce quickly, let loose with a high rate of mutations, and recombine the subunits of the instructions for specific antibodies, almost like replacing a few cards of a hand in a life-and-death game of poker. The newly invented antibodies are tested in the centers of lymph nodes to see if they work well. Almost none of them do. To nip the proliferation of bad ideas in the bud, the cells that came up with them are killed. But a few of the antibodies are better. The cells that produced them are allowed to live and encouraged to grow. The defense against the invaders improves. When the population of invaders is destroyed, most of the specialists that were marshaled are no longer needed and commit suicide; a few are left as memory cells to respond quickly should the same invader return.

This description, which only hints at the underlying complexity of immunological defenses, makes it sound as though these cells of the immune system have brains. Indeed, they act as if they had brains, but they do not. All the interactions are orchestrated by chemical signals and chemical responses. Chemical messages are released from one cell to another, across membranes, and to different parts within the cell to control the cell's mechanical responses and its reading of genetic instructions. Individually, these defense cells do not have brains, but taken collectively, their ability to process and communicate information bears a surprising resemblance to the brain.

This magnificent immune system fails us sometimes because all this identification, communication, mobilization, reproduction, and fine-tuning takes time. While the immune system is performing these activities, the pathogens are busy reproducing and invading. Once the invasion is identified and the immune system mobilized, the outcome depends on how quickly the immune system can make up for lost time. When a novel pathogen is encountered, the time from the initial invasion to immunological control typically takes about a week. If the same invader is encountered by the person a month or a year later, the response time is much shorter—typically a couple of days, thanks to the quick responses of the memory cells. It is so fast and effective that the person may not even be aware of the invasion. With the pathogen-host evolutionary arms race running neck and neck, the shorter time makes all the difference—the difference between illness and health, and sometimes between survival and death.

USING OUR OTHER BRAIN

Vaccines are simply a tweak that shifts the immune system response from the longer delay to the shorter delay. The tweaking of the immune system with vaccines has so strongly aided the immune system in this conflict that it has eradicated one scourge—smallpox—and virtually eradicated several others, such as polio, measles, and diphtheria, from large regions of the planet. We take pride in our vaccines, but really the vaccines are the simplest part of the defense. A vaccine is a mug shot of a criminal sent to a police station before the criminal is encountered. Sending the mug shot can be terribly important because it allows the police force to recognize and respond quickly to the real criminal. Merely sending the mug shot, though, is far less complex and difficult than tracking down, apprehending, and incarcerating the criminal.

Put bluntly, medicine's successes at vaccination and antibiotic treatment are trivial accomplishments relative to natural selection's success at generating the immune system. Recognizing this fact has important repercussions for the long-term control of infectious diseases. We will probably obtain much better disease control by figuring out how to further tweak the immune system and capitalize on its vastly superior abilities than by relying on some human invention such as new antimicrobials (antibiotics, antivirals, or antiprotozoal agents). Antimicrobials are useful, but the problems they are good at solving differ from the problems vaccines are good at solving. Antimicrobials are not particularly suited to controlling or eradicating disease at the population level. Rather they are good at helping a patient who needs to control an infection now. Vaccines are effective at protecting individuals from becoming infected, at controlling the spread of disease through populations, and sometimes at eradicating a disease from a population. The blurring of this distinction between the roles of antimicrobials and vaccines has hampered the control of disease and exacerbated the dilemma of antibiotic resistance.

I am confident that we have only begun to capitalize on the immune system, though I doubt that we will soon be able to administer safely and effectively the chemical messengers used by the immune system to turn a particular response up or down. Some scientists hope that a more effective immune response could be generated by increasing or decreasing some of these chemical messengers, like turning the volume control on a radio up or down. I expect that the immune system is much too weblike to do that. Radios were engineered to be controlled by a listener. They therefore have controls for particular properties, such as volume, treble, and bass, controls that are well suited for adjustment by the fingers of the listener. The immune system, like the brain, was engineered by natural selection to be a self-controlling unit. It therefore does not have controls that allow an outside user to turn one attribute at a time up or down. If we increased a chemical messenger to try to improve immune function, we would probably cause many unforeseen effects as the immediate response affected other parts of the immunological web.

We can use the brain analogy to get a sense of what can be done with the immune system. If we start stimulating a neuron here and a neuron there, we are unlikely to improve the brain's ability to function. Attempts to actually change the circuitry have ranged along the spectrum from very crude to moderately crude solutions; they may solve one problem but introduce others. One need only mention the gruesome failures of prefrontal lobotomy and electroshock therapy. A hands-on approach to the brain works best when some neuronal damage causes a shortage of a specific chemical, the effects of which can then be partially ameliorated by supplementing the shortage with the same chemical or a similar one. The use of L-dopa for Parkinson's disease is this kind of solution; the L-dopa compensates for the shortage of the chemical transmitter dopamine. But even in this case the solution is generally only partially effective. We are headed for the same kinds of disappointments if we try to treat the immune system like a radio that can have its components adjusted, replaced, or removed.

We can effectively improve the brain's functioning, however, by making use of the brain's own already wired abilities to improve—we call it learning, or training. Vaccines similarly make use of the immune system's already wired abilities to control microbes. Vaccines teach and train by providing information to the immune system, information that we humans know sooner than the immune system knows, information about what pathogens are out there ready to invade before they invade. That kind of tweaking has worked marvelously in the past, and it will undoubtedly do so in the near future. Vaccines teach the immune system, but good teachers know they cannot just present the information and leave it to the students to sort it out. Some students will be able to grasp an idea because of their background and inherent abilities, and others will not. A given student will grasp some ideas very well and other ideas poorly. The information that goes out will influence what is absorbed. If a teacher sends it out at too high an intensity, the student can be overwhelmed, and the entire subject can become frustrating and destroy the student's interest, no matter how inherently interesting and manageable the subject. Medicine needs to figure out how to better teach the immune system, rather than trying to adjust and reconstruct it.

The immune system is an information-processing system that is analogous to the brain, but with different input and a different function. The brain deals with information from the outside world that comes in through sight, hearing, smell, taste, and touch and integrates it in the form of ideas and concepts; the brain's goal is to interact more adeptly with relatively large organisms, those that we recognize as predators, competitors, and food sources. The immune system takes in information about small invading organisms, integrates it, and mobilizes it with the goal of interacting more adeptly with the microscopic environment. The goal tends to be more restricted, being largely the microscopic analog of avoiding predators and ectoparasites. But the task is more diffuse, more akin to controlling entire police forces or entire military operations than to altering the behavior of one person or one predator.

We rarely think of the immune system as an analog of the brain—as another decision-making system in our bodies—probably because the immune system does not generate sensations. We are therefore oblivious to its information processing and decision making. But it is there, and it integrates information and actions in very complex ways in each of us at every moment. We need not let our senses fool us into overestimating the importance and complexity of one system relative to another just because the one generates sensations and the other does not.

What options for tweaking are there besides vaccines? The brain function of my students can be overloaded with input. Immunological processing can be overloaded by pathogens if the dosage is too high, the harmfulness of the pathogen too great, or the route of entry too direct. One key to a truly preventive medicine will be to intervene in ways that keep the threat of pathogens well within the ability of the immune system to deal with them. We can keep the immune system from being overwhelmed, for example, by adjusting things so that pathogens entering the body have a combination of low dosage and low inherent harmfulness, but this is a formidable task.

The good news, which will be developed in the remaining chapters of this book, is that evolutionary principles offer new directions that will probably allow disease prevention programs to make better use of our immune systems by better controlling the evolution of their microbial adversaries. In short, there are ways to make vaccines, use antimicrobials, and improve hygiene so that control of pathogens by the immune system is much more manageable. Often these new methods capitalize on the immune system's ability to mediate competition. If we intervene in ways that expose the immune system to milder organisms that are circulating in the population before they see the more harmful competitors, we can get the immune system to be better prepared for the more dangerous organisms, very much as if it had been vaccinated against the harmful strains.

I will return to these matters fully in the third part of this book, but first we need to grasp the scope of infectious disease. How can we identify the infectious plagues of the present? Do pathogens cause most cancer, heart disease, Alzheimer's, and other ailments that we have been tolerating as grim mysteries? What conceptions of disease are responsible for the current stagnation in the control of infectious diseases?

PART II

INFECTIOUS THREATS NOW

On Guard Against the Acute—

Blindsided by the Chronic

CHAPTER FIVE

The Endless War

THE PLAN SEEMED REASONABLE TO GENERALS MOVING FIGURINES over maps. The British artillery would pummel the German trenches. The infantry would then storm the trenches and mop up the remains of the German troops. The plan might have worked nicely in the Crimean War, over a half century earlier. But at the Western Front of World War I, it turned out to be an effective way to kill and maim your own soldiers at a rate of thousands per hour. The German soldiers took refuge in bunkers deep below their trenches during the bombardment. As soon as the bombardment ended, they were ready at their machine guns. A wall of bullets met the British soldiers as they tried to run the few hundred yards across no-man's-land. It was a new incarnation of an old mistake: the fighting-the-last-war problem.

Those in charge of health programs often use a war metaphor to communicate their goals and expectations. They are, however, selective. They do not like to compare their programs to the stalemate at the Western Front in the First World War. The more decisive battles of the Second World War provide the preferred model. U.S. political leaders of the 1950s rallied the population for a World War II–style health campaign with the heady success of that war still fresh in everyone's minds, and with powerful new weapons at hand—penicillin for bacteria, DDT for mosquitoes, chloroquine for malarial parasites, and improved technologies for viral isolation and vaccination. President

Dwight D. Eisenhower called for "the unconditional surrender of microbes." Secretary of State George C. Marshall foresaw the "imminent conquest of disease." Senator John F. Kennedy predicted that "children born in the next decade would no longer face the ancient scourge of pestilence." Medical leaders provided the expert opinion that justified such optimism. A decade after JFK's prediction, the surgeon general spoke of closing the book on infectious disease. In the 1972 version of their classic book *Natural History of Infectious Disease,* the Nobel laureate Macfarlane Burnet and his coauthor, David White, predicted that "the future of infectious diseases will be very dull." The children to whom JFK referred had reached their twenties when the AIDS pandemic erupted, when cervical cancer was recognized as an infectious disease, when hospital-acquired infections were recognized as the tenth leading cause of death in the United States, and when in vitro fertilization came into the vernacular as a way of coping with the epidemic of infertility caused by flourishing venereal diseases.

The victory-in-war metaphor is not just a post–World War II blip on the public health radar screen. It has been invoked, though with less bravado, since the establishment of the germ theory at the end of the nineteenth century. Before the can-do postwar decades, mainstream medicine took for granted that infectious diseases would be conquered. In the midst of World War II, the Yale professor of medicine Charles-Edward Amory Winslow wrote a history of medicine that he matter-of-factly called *The Conquest of Disease*; Winslow provided no defense of his proposition, and no discussion of the possibility or probability of conquering epidemic disease. It was manifest destiny. He concluded his book with the statement that "the practical application of the principles developed by a series of clear thinkers and brilliant investigators . . . has forever banished from the earth the major plagues and pestilences of the past." In his zeal for declaring victory, Winslow seems to have overlooked some diseases. Falciparum malaria, for example, one of the most harmful diseases in human history, had barely been nudged back in 1943 when Winslow was writing. The rich-country bias, which seems to be at the root of Winslow's hubris, persisted among leading health scientists right up to the late 1970s, when the AIDS epidemic and antibiotic resistance gave us a reality check. Burnet and White showed similar exuberance: "Young people today have had almost no experience of serious infectious disease." The main problem with this statement was that it didn't apply to two thirds of the planet's young people—those who were living in poor countries at the time.

Though today's leaders in disease control have been humbled by the limited success at global conquest, they still visualize their approach to infectious disease in terms of warfare. When the science journalist Wendy Orent asked John W. Huggins, an expert on monkey pox, about the possible evolutionary outcomes of monkey pox transmission among humans, Huggins responded, "I don't think like an evolutionary biologist. I just want to find a drug for these bugs and kill them." An evolutionary biologist would be disappointed in Huggins's first sentence and take aim at the word *just* in the second. The twentieth century has left us with a clear message: it will not be enough to just try to find a drug to use as a lethal weapon against the pathogens of our future. We must understand why they are the way they are, and use this understanding to manage their evolution.

THE WAYS WE HAVE FOUGHT

Why has the war metaphor so dominated our approach to infectious disease? One reason is that the metaphor has been successful in some arenas. We humans rally to the defense against a common enemy, and this support led to the control of many damaging infectious diseases during the first half of the twentieth century. Still, the only case of complete victory—eradication of the enemy—has been the smallpox campaign, and that vaccine was no high-tech wonder. William Jenner took pus from a sore on a cow and used it to inoculate those whom he hoped to protect from smallpox. The primary technological advance for vaccine development involved growing the still mysterious vaccinia virus in cows to amplify the virus, and then scraping the infected cow tissue to get the material for the vaccine.

If we allow the definition of war to include organized attacks on populations of other species, the eradication of smallpox during the 1970s was literally a war, the most cost-effective antimicrobial war that has ever been fought. It was accomplished in just over a decade with a modest price tag of $300 million. The army comprised a few hundred officers (physicians and epidemiologists) and a few thousand foot soldiers (assistants working in the lab and outside community). In 1977 they rooted out the smallpox virus from its last refugia: Bangladesh and Somalia. Their efforts were heroic but their enemy was vulnerable. As microbial targets go, the smallpox virus was a clay pigeon. Its ability to change form was so limited that vaccination with a different virus of the same evolutionary group, the vaccinia virus, could trigger an immunity that would knock out the entire spectrum of smallpox viruses.

To win wars, social values must be compromised. One of the first to be diminished is the value of human life, because war involves the killing of people. Freedom is another casualty. Imagine what would have happened if each soldier in the trenches of the Somme or Galipoli had been given a free choice to go back to his family without penalty or to die in a fruitless rush against the enemy trenches. How many would have stayed?

The actions taken in microbial wars need not be as severe, but it is wishful thinking to expect that we can win these wars while adhering to all our peacetime values. Administrative agencies such as the World Health Organization set up guidelines to help preserve such values. Yet even in the war against smallpox—the most lopsidedly winnable of our microbe wars and the most illustrious victory of the World Health Organization—these guidelines had to be violated. One of the commanders in this war, Isao Arita, later confided, "If we hadn't broken every single WHO rule many times over, we would never have defeated smallpox. Never."

Part of the problem is that the need to break these rules increases the closer one gets to eradication. Some people are willing to receive a vaccine, others are obstinately opposed. The early positive effects of vaccination provide an inflated sense of success because the success is enhanced by those who choose to cooperate. The completion of the campaigns often requires that the last holdouts are persuaded to cooperate or forced to submit. All eradication campaigns suffer from this Achilles' heel. Some people want to have their houses fumigated, others resist. Some prefer to have their contagious illness treated, others refuse. If we are to win some of the more winnable wars—those against measles and polio, for example—we may need to violate more rules than have been violated to date.

Vaccines were the first great strategic success in the war against infectious disease, but the return on the investment has steadily declined as increasing amounts of money have been invested in this option. The most successful vaccination campaign, the eradication of smallpox, is attributable to an eighteenth-century innovation, the vaccinia vaccine. It cost almost nothing and swept aside one of humankind's worst infectious adversaries. The great vaccination efforts of the first half of the twentieth century were cheap and marvelously successful. Vaccines against diphtheria and tetanus were developed with less money than is now sometimes awarded in one grant from the National Institutes of Health.

Vaccines against pertussis and polio were developed during the 1950s; they were successful but came with greater financial cost and more collateral damage than the smallpox program. Some of the viruses in live polio vaccines reverted to neurovirulence, causing paralysis in a small number of those vaccinated. Moreover, vaccine viruses proved infectious for those in contact with the vaccine recipients. This secondary transmission of vaccine virus increased the protection of the population and was therefore considered a bonus so long as the viruses did not revert to neurovirulence. But polioviruses are particularly prone to mutations. The live viruses in the polio vaccines were like unseasoned soldiers who could not be trusted to protect without shooting innocent civilians. The threat of reversion to neurovirulence has been recognized since it was encountered in Philadelphia in 1935 when an experimental polio vaccine was administered to several thousand children. Viruses from an attenuated live vaccine reverted to neurovirulence, killing five of the nine children who contracted polio from it. Reversions to neurovirulence have occurred sporadically since then and have made vaccine researchers apprehensive; they are hoping that the vaccine viruses left as a legacy of the live poliovirus vaccination programs will not develop similar reversions to virulence, generating new polio outbreaks where wild polio has been eradicated. A more sinister problem was recognized by Michele Carbone in the early 1990s. Some polio vaccine batches were contaminated with a monkey virus, SV40, which now appears to be causing cancers of the brain and lungs in a small percentage of those who were vaccinated in the 1950s and early 1960s.

Live vaccines against measles and mumps were developed in the 1960s, and a few more vaccines were added during the last three decades of the twentieth century: vaccines against hepatitis viruses A and B, *Haemophilus influenzae,* and the perennial ones against influenza. The effectiveness of these vaccines has varied from the magnificent suppression of encephalitis by the *Haemophilus* vaccine to the erratic performance of the influenza vaccines. But the important point is that these successes came from increasing expenditures in vaccine research. The cost-effectiveness of vaccine development has eroded because of the many failures, such as the efforts to develop vaccines against malaria and AIDS. Vaccine development certainly has a place in our future, but their use as weapons to destroy infectious adversaries is becoming more costly and generally less effective. We will need to develop and use vaccines more cleverly to improve their performance.

The second great strategy is hygienic improvements. No other single intervention in the history of medicine has saved as many lives and reduced as much suffering as the provisioning of uncontaminated water, which is necessary to curb the great plagues of diarrheal disease. Before cities cleaned up their water supplies, about one out of every five residents would die of diarrheal diseases, particularly typhoid fever, dysentery, and cholera. Provisioning of clean water dropped this death toll virtually to zero. Hygienic improvements in hospitals during the same period transformed hospitals from institutions that served as the last stop for the living to places where people had a good chance of being cured, particularly after antibiotics were introduced.

We can think of hygienic improvements as the strafing of enemy forces as they try to cross bridges, the point being that enemy forces are stopped by disinfecting, filtering, or washing them away while they are en route to their target. Those with a less militaristic disposition can equate hygienic improvements to the Marshall Plan, which rebuilt postwar Europe. Infrastructural improvements that encourage safe drinking water, sewage disposal, hand washing, disinfection of hospitals, and use of gloves allow people to improve their lives by reducing the ravages of microbial warfare. Whatever the metaphor, the overriding argument is that hygienic improvements work and are relatively inexpensive. We relax them at our own peril.

The third great strategy against infection is the use of antimicrobials. When antibiotics were introduced, some hoped that they might put an end to bacterial diseases. The track record, as of the mid-1940s, bolstered this lofty expectation, at least for those who did not bother to consider evolution. Those who did, however, sounded an alarm almost from the beginning of the antibiotic era. The discoverer of penicillin, Alexander Fleming, warned in the late 1940s that antibiotics might soon lose their effectiveness through the evolution of antibiotic resistance. Another Nobel laureate, Joshua Lederberg, voiced a similar warning during the early 1950s. But the alarm went largely unheeded.

Although antibiotics are a marvelous solution for the individual patient, they are a poor way to control disease in a population. Their flaw stems from an ethical dilemma: what is best for the individual patient may often be at odds with what is best in the long term for the society. The individual patient usually benefits by taking antibiotics. The society pays a price over the long term because the more antibiotics are used, the stronger are the selective forces favoring the evolution of antibiotic resistance. Societies might reduce this price if they were able to restrict antibiotic usage, but they have few options. If the restrictions are voluntary, a physician who does not treat a sick patient with an available drug will likely lose the patient—if not to death, then to another physician. If the restrictions are mandatory, their enforcement would undoubtedly lead to a black market in effective antibiotics. Even if societies had the power to enact such restrictions, ethical considerations might cause them to choose not to exert it.

Should the possible long-term interests of the society as a whole restrict the access of individuals who need an antibiotic now? The antibiotic might not even be necessary in the future if other antibiotics were developed in the meantime. There is no easy way out of this dilemma. The best long-term solution is undoubtedly to reduce the need for antibiotics rather than to restrict their availability. But this solution requires clear and clever thinking about evolutionary processes, something that has been in short supply in the health sciences throughout the past two centuries.

Still, for all the negative press antibiotics have received in the context of antibiotic resistance, they remain one of the three great achievements in the fight against infectious disease. Many of the difficulties that are encountered with antibiotics result from a failure to distinguish between purposes for which they are particularly suited and those for which they are not. Antibiotics are an excellent weapon for destroying the microbes that are on a rampage in a sick patient. They are a poor weapon for controlling the spread of infectious agents through a population—when they are put to this use, problems of antibiotic resistance are sure to be exacerbated. Hygienic improvements and vaccination are the better weapons for this goal.

This rule of thumb sounds simple, but it is frequently violated. It is violated when hospital staff decide to treat wards prophylactically with antibiotics instead of maintaining the high hygienic standards that prevent infections. It is violated when antibiotics are offered to poor countries as a cheap solution to diarrheal diseases instead of making the more expensive improvements in water supply and sewage disposal. It is violated when tuberculosis control policies rely on seeking out and treating the infected individuals in a city rather than improving housing quality, airflow, and nutrition in the hotbeds of tuberculosis transmission.

THE PAYOFF FROM MEDICINE

Medical research is consuming far more funds than it was when the health sciences were making breathtaking advancements in the quality and quantity of human life. We spend billions of tax dollars, for instance, on the budget of the National Institutes of Health every year. When it comes to the control of disease, medicine gave us much less for much more money as the twentieth century progressed. It is true to some extent that the problems faced now are harder to solve, but it is also true that medical research is stuck in a rut. We can be confident only that the problems of the present are harder to solve using the approaches of the past. Little in the control of disease during the last half of the twentieth century involved fundamentally new approaches, in the same way that antibiotics are fundamentally different from vaccination, which is fundamentally different from hygienic improvements, which is fundamentally different from surgical removal of diseased tissue. Rather we are recycling each of these solutions. The major advances during this period were vaccines and antibiotics. The payoffs from vaccination dwindled from the successes against polio, measles, whooping cough, and mumps during the 1950s and 1960s to the more sporadic successes of the subsequent three decades. The generation of new antibiotics struggled to stay ahead of antibiotic resistance. The antibiotic strategy was broadened to encompass antiviral and antiprotozoal drugs, but these applications never generated the magic bullets that were anticipated from the successes of the 1940s. The positive effects of antivirals on AIDS, and antimalarials on malaria, never came close to the effects of penicillin on streptococcal pneumonia. When confronted with problems such as influenza, AIDS, and malaria, researchers kept applying the solutions that worked for other diseases, hoping that improved technology would improve the results of these marginally effective strategies. They did not use technology to generate any fundamentally different approaches.

In spite of this declining return on investment, one fact stands out. Though there has been much investment in studies of genetic causation of disease, the most significant returns were generated on the investments in studies of infectious diseases. Polio, measles, hepatitis, liver cancer, ulcers, and cervical cancer were all controlled or shown to be controllable during the last half of the twentieth century through the control of infectious agents. This track record provides a sense of how to invest in our future: focus on the germs.

The Human Genome Project serves as a test case. Many medical experts believe that the Human Genome Project will accomplish much of what the last half century of medical genetics failed to do: identify the genetic causes of the major diseases so that the harm from these diseases can be ameliorated. The genome project will surely illuminate genetic causes of disease, but evolutionary principles suggest that the most important disease-causing genes will belong not to the humans but to the pathogens. Human genes will become relevant not so much because they cause disease but rather because they protect or fail to protect against pathogen genes.

Specifically, analysis of the human genome should foster the discovery of genetic sequences that seem out of place in the human genome. These stretches will seem out of place because they *are* out of place, being viral rather than human. By facilitating an improved understanding of infectious agents in the human genome, these studies will help unmask potential agents of human diseases.

To understand fully why we are so compelled to portray control of infectious diseases as warfare, we need to look inward rather than outward. The human mind did most of its evolution as our ancestors were gathering food, hunting prey, avoiding being hunted as prey, and fighting and negotiating with other humans. This kind of problem-solving required memory, ingenuity, and the dexterous use of tools and weapons. All these activities can be well conducted by minds that model the environment as being composed of building blocks to be destroyed, avoided, or taken and used. Things that cause harm are cast as enemies, often personified and vilified. If the goal is to eliminate the harmful condition or offending process, casting it as an enemy helps us to visualize it and its elimination. To recognize this tendency, we need look no further than the "wars" in which we engage: the War on Poverty, the War on Cancer, the battle between good and evil. In none of these examples is the enemy really an organism that is killed in battle. The War on Poverty was a government investment to help poor people (and perhaps to solidify political power). The War on Cancer was a research effort, and the battle between good and evil is an effort to alter people's beliefs and make them act in accordance with some model of behavior. But these bland descriptions of goals do not generate a picture that engages the mind with the cause at hand. The war metaphor does. In addition, the war metaphor provides an engaging mental picture for the disease eradicators. It casts the disease as an enemy and allows them to use the medical versions of spears, clubs, and fire—antibiotics, disinfectants, and vaccines—to destroy or ward off the enemy.

The purpose of this brief excursion into the evolution of human behavior is to suggest that we must be on guard to recognize the errors into which the human mind might easily slip. One error might be a failure to recognize opportunities for controlling disease. When we are faced with a disease organism, we may just want to kill it like an enemy rather than trying to control its biology in a more beneficial way. Another error might involve a more global view. Throughout our evolutionary history human enemies must have been invading from the outside. Mental imagery of invasion may have been favored by natural selection because it would have allowed our ancestors to plan for the attack. But when applied to disease control, the imagery may introduce biases. Invasion imagery was certainly appropriate for the disease epidemics that invaded the New World from the Old World, and one part of the Old World from another through the nineteenth century. It helps explain why at the close of the fifteenth century, the Italians called syphilis the French disease, and the French called syphilis the Italian disease. It also helps explain why influenza epidemics typically bear names such as Hong Kong flu and Spanish flu (at least outside Hong Kong and Spain). Influenza epidemics certainly do cross national borders, as did the AIDS pandemic. This border-crossing ability of infectious plagues has led many to believe that the greatest infectious threats will come from the outside. But we need to remember that our expectations are biased by minds that tend to visualize threats as invaders coming from beyond our borders. Perhaps the experts are looking in the wrong place. Perhaps the most menacing infectious adversaries are already here.

CHAPTER SIX

Where Are They Coming From?

IN HIS BOOK *GUNS, GERMS AND STEEL*, JARED DIAMOND writes of North America as a melting pot for the world's pathogens. It certainly is a melting pot for pathogens, just as it is still a melting pot for humans. But is the gathering of foreign pathogens in the melting pot of North America the real problem? Are the most important threats to North America coming from distant places? There certainly is no shortage of pathogens crossing international borders. A new scare surfaces almost every year. In the summer and early fall of 1999 it was the West Nile virus, which killed seven people, some horses, and several thousand crows along the eastern seaboard of the United States. Then the outbreak fizzled. Even when a full-fledged vector-borne pathogen like the dengue virus or the malaria protozoan gets into the United States, it peters out on its own. If these masters of vector-borne transmission cannot make a go of it in humans, what chance does a West Nile virus have? It is out of its element in humans. It cannot be transmitted from human to mosquito to human even in the best of situations. It attracts attention mainly because it is exotic.

People tend to overestimate the risk of accidents that are rare but sensational—the airplane crashes—relative to the much more dangerous but commonplace accidents—the car crashes. So too we tend to overestimate the danger from the unusual when it is lethal. But statistics can correct our mistakes. People who look at the actual calculations of the probability of dying per mile traveled in an airplane can see that it is vastly less than the probability of dying per mile traveled in a car and adjust their worries accordingly. The same thing can be done with infectious diseases, though people have to spend a bit more effort in educating themselves about the diseases with known microbial causes, the microbes that cause them, and the diseases that may be caused by unknown microbes. If people do not do this extra legwork, the government and its control agencies will continue to go down the path of least resistance. If the public is more frightened about Ebola, hantavirus, Legionnaires' disease, and West Nile virus than about the probable infectious causes of chronic diseases like cancer, stroke, heart attack, multiple sclerosis, and Alzheimer's, then the public will get overfunding of low-level infectious threats and underfunding of high-level infectious threats.

Because the AIDS virus originated in sub-Saharan Africa, the AIDS pandemic has fixed worried attention on what might be out there ready to emerge from steamy African jungles or urban Asian swarms, just a plane ride away from the hubs of Euro-American civilization. Many leading experts have called for interventions to block this emergence. Jonathan Mann, the first head of the World Health Organization's AIDS program, called for a worldwide early warning system "to detect quickly the eruption of new diseases or the unusual spread of old diseases." It sounds like a good idea on the face of it, but how effective would it be? The history of disease emergence provides some clues.

Even superficial digging unearths many examples of diseases that immigrated into prosperous Western countries from poorer ones: yellow fever, repeated outbreaks of influenza, Asiatic cholera, black plague, malaria, and of course, AIDS. But there is something telling in this list, or any other representative list one might construct. With the exception of AIDS, the diseases first made the trip to Euro-American populations many decades or even centuries ago.

Their global mobility depended on two critical factors: the group size of the traveling humans and the speed of travel. Human travel needed to be sufficiently rapid to allow disease organisms to be maintained within the group of travelers. Meeting this requirement is more tricky than it sounds because the human immune system is so good at subduing pathogens. Within a few weeks pathogens are usually eradicated from the body or so decimated that they have little chance of transmission. Pathogens that had mastered the trick of prolonging infectiousness were probably transported between the Mediterranean and south and central Asia from the earliest days of intercontinental trade. Pathogens that had not mastered the trick would have to wait for improvements in transportation technology to move them throughout the civilized world.

But they would not have to wait for jumbo jets. Caravans and horses were sufficient for diseases such as the black plague to make their way from central Asia to southern Europe nearly two thousand years ago. For other diseases, the movement of big boats permitted travel over expansive tracts of humanless terrain. In the early nineteenth century Asiatic cholera finally made the passage from its south Asian homeland to Europe and then did so repeatedly, causing waves of turbulence in Europe and America until water supplies were protected. Boats filled with dozens of people may have been necessary to permit its passage because the cholera organism can rarely withstand a person's defenses for more than a couple of weeks. Yellow fever similarly was barred from the Americas until ships could act as microcosms, complete with tiny pools of freshwater in which the mosquito vectors could breed. The African mosquito *Aedes aegypti* colonized the boats and eventually the urban centers of the New World, thus cycling the virus in people both during and after the transatlantic voyage. Jet travel is overkill, by about five hundred miles per hour.

The emergence of new disease organisms from human enclaves was once a great danger that often materialized into great devastation.

Plague carried off one third of the European population in the fourteenth century. Yellow fever created more sporadic havoc, biting out sizable chunks of cities such as Memphis and Philadelphia in the seventeenth and eighteenth centuries. Falciparum malaria was present in ancient Egypt and probably hitched rides throughout the Mediterranean as humans crisscrossed this ancient marine thoroughfare. The list could go on. From ancient Egypt to nineteenth-century London, new diseases were rolling, riding, sailing, and steaming their way into populations around the world.

But the time when these concerns were most warranted is over. The last big hurrah of world exploration by human pathogens corresponded to the period of human exploration and colonization. If Christopher Columbus had had a high-tech surveillance system at his disposal, the mind-set of a public health expert, and the ability to influence public policy, *he* would have been in a position to do some major disease prevention. This last great expansion of infectious diseases gets less airplay than the current threat of emerging diseases because humans are self-centered beasts, being more attentive to the minor dangers that threaten us today than to the devastating problems we imposed on others in the past. The most devastating new diseases in recorded history were those transmitted *from* rather than *to* Europeans. American Indians, Polynesians, and aboriginal Australians did not write our history books. There were not many of them who were in a position to do so, partly because there were not that many of them. Most had been destroyed by European diseases such as smallpox and measles.

The history of the twentieth century indicates that the greatest threat to Europe and North America from African and Asian diseases has already come and gone. What nasty new epidemics spread to Europe and North America from Africa and Asia during the twentieth century? Only AIDS. Even the most notorious pandemic of influenza, which has given some notoriety to east Asia as a virological melting pot, cannot be attributed to Asia. As I noted in the first chapter, the best reconstruction of evidence traces the origins of the exceptionally virulent 1918 viruses not to Asia but to the western front of Europe, where the conditions of trench warfare and the transport of severely ill patients apparently favored evolutionary increases in the virulence of the influenza viruses that were circulating there.

The greatest threat to humans from pathogens is from pathogens that are already widespread in humans. True, some pathogens can cross into humans from other species in faraway places, but we now have a good sense of which ones can generate a grave threat. We have tracked this drama with yellow fever and influenza repeatedly. Long before the AIDS pandemic, most of the dangerous pathogens from African jungles or Asian multitudes had amply demonstrated their particular horrors.

Certainly, imported diseases have recently caused problems for small numbers of people in wealthy countries. Increased travel between rich and poor countries exposes travelers to the unsolved problems of the poorer countries. The blood of visitors to sub-Saharan Africa has a good chance of acquiring the malaria parasites that annually kill a million people in poor countries. The intestinal tracts of travelers to almost any poor country have a good chance of becoming hosts to the waterborne viruses, bacteria, and protozoa that annually kill millions of people by causing diarrhea.

Every day, travelers bring these pathogens into JFK and Los Angeles International Airport, fresh from the poorer countries. These infected travelers may serve as the source for a few additional infections that arise within our borders, but by and large these chains of infection peter out, largely because the infrastructure of rich countries inhibits their spread.

Consider dengue. The dengue virus is continually slipping across the Texas border from Mexico, but it does not spread within the United States, probably because people on the Texas side spend most of their time in mosquito-proof houses, cars, and workplaces. By keeping people inside mosquito-proof structures, computers and televisions may help curb the spread—the virus becomes stranded like a traveler in an airport in which all flights have been canceled.

Similar evidence comes from a different disease at the other end of the border between the United States and Mexico. Bloody diarrhea killed thousands of people from Mexico south through Guatemala during the late 1960s and early 1970s. But besides a few tourists who brought home a life-threatening infection, the epidemic hardly affected U.S. residents. Some people were infected by those who brought the lethal bacteria north of the Mexican border, but as with dengue, the chain of infection petered out on its own, this time probably because even the poorest U.S. citizens tend to have good access to drinking water that is free from fecal contamination.

The medical history lesson of the twentieth century is that people in rich countries are protected against many of the worst infectious diseases because of infrastructure. Mosquito-proof housing, sewage disposal, basic hygiene, and clean water protect us better than any surveillance system. Countries without such protective infrastructure are vulnerable. But would an early warning system reduce their vulnerability?

The recent Latin American experience with cholera provides some insight. Remembering the nineteenth-century experience—"*en los tiempos del cólera,*" as Gabriel García Márquez put it—Latin America braced for cholera's return during the 1970s. The threat seemed imminent as a new cholera pandemic flowed northward from Indonesia in the 1960s, to India in the west and to China and the Philippines in the east. By the early 1970s it had spread through the Middle East and into Africa. The front spread wherever food and water were poorly protected from fecal contamination. By the mid-1970s it had reached the western countries of sub-Saharan Africa. Latin America now seemed particularly vulnerable because the infrastructure that could protect the richer countries was absent from most of Latin America. The Atlantic Ocean lay between South America and cholera-ridden West Africa, but the Atlantic was an insufficient barrier in the nineteenth century; it could not be trusted to provide a barrier in the twentieth century, when it could be traversed by jumbo jets in a few hours. Yet the pandemic did not spread into Latin America in the 1970s, nor in the 1980s.

It did in the 1990s. It entered Latin America by the back door, arriving in Lima, Peru, in January 1991. Within a year it had spread throughout South and Central America, and it shows no sign of leaving. Its toll during the 1990s was over a million cases and tens of thousands of deaths.

The poor countries did not need an early warning system for cholera. They had decades of lead time. Rather, they needed the infrastructure of the rich countries—but not the entire infrastructure. They needed then, and most still need, the two keys to protection: clean water and adequate waste disposal. No country with these two assets has suffered much from cholera. Even in Latin America this generalization holds. Chile, for example, ranks relatively high in water quality and waste disposal. Within two months after the first cholera cases showed up in Lima, the disease entered Santiago, Chile. Then again about a year later, it entered Chile's northern provinces. During the first few years of cholera's visitation, Peru reported over a hundred thousand cases. Chile, Peru's next-door neighbor, reported under a hundred. In 1994, when Peru reported about 25,000 cases, Chile reported one.

THEY'RE HERE

Some infections will not peter out on their own even with all the infrastructure money can buy: infections transmitted by sex. But almost all the venereal diseases that are important in one geographic region have already spread globally. This global spread happened long ago because sexually transmitted pathogens tend to have prolonged periods of infectiousness, and people the world over are having sex. People infected with sexually transmitted organisms can carry them across oceans and mountains and around the world, whenever the people themselves can make such voyages.

Could an early warning system work against these truly dangerous globe-trotting pathogens? Some experts imagine massive investments in surveillance and interdiction, but that proposal sounds disturbingly like the efforts directed against drug smuggling. Surely it would be an act of desperation. Surveillance and interdiction might provide an effective barrier against pathogens that pose little threat of spread, such as the Ebola virus, but that would be like building two cages around a dangerous animal when one cage would do. How could such containment work for the more serious threats, such as a new sexually transmitted pathogen, a new AIDS virus? Such pathogens are generally asymptomatic or cryptically symptomatic during much of their infectiousness. Are we going to test each of the millions of international travelers for venereal infection and make them wait with immigration services until the results of their tests are available? The complete failure to control the spread of HIV into new countries during the last decade in spite of the clearest signs of the impending disaster reveals how hopeless such an approach would be.

Experts on AIDS are currently arguing about whether HIV arose naturally through transfer from chimpanzees to humans, or whether medical activities, such as vaccination or reuse of contaminated syringes, played a critical role. If HIV arose naturally, it serves as a warning of the kinds of things that could happen naturally in the future. Yet even if HIV did arise from a chimpanzee virus without any help from medicine, there is a bright side: few new AIDS-like pathogens are likely to invade from some isolated group of humans because there are hardly any isolated groups of humans left on the planet. The recognition of the exaggerated threat from far-off places does come with a few caveats, however. One caveat involves the historical recognition that germs have entered human populations primarily from two sources. The first source is domesticated animals. Unless we start introducing pathogens from domesticated animals in new ways—such as through the transplanting of their organs—the threat from domesticated animals probably holds few new surprises. The second source is other primates. Our biological machinery has diverged evolutionarily from that of other primates more recently than it has from more distantly related species. If a germ enters a human from another primate, especially from another ape, the barriers to setting up shop will be relatively low. At the time of the transfer, the germ's characteristics will have evolved to take advantage of that other primate; but because that primate is more similar biochemically to humans than, say, lions, we can expect a greater proportion of these transfers to take hold.

The process is much like the differential settlement of North America. Immigrants did not settle the continent randomly but rather settled in places where the climate and topography was similar to their homeland. Scandinavians tended to settle in Minnesota, and Germans in Wisconsin. Those from the British Isles settled more heavily along the Atlantic coast. To some extent these differences depended on what was available, but that cannot be the whole story. Who would be more likely to say that Minnesota winters are not so harsh—an immigrant from England or an immigrant from Norway? Germs, of course, do not have the option of forethought; still, one can ask whether a germ that enters a human from a chimpanzee will find life in humans more harsh than would a germ from a lion. It's no surprise that although African cats and African primates have retroviruses, we humans have gotten our retroviruses, the HIVs and the HTLVs, from other primates rather than from cats.

It is not just the ability to infect and grow in humans that is important. It is also the ability to get out of a human and into the next human. Many germs have passed the first test but failed the second. The track record indicates that here, too, germs from other primates are more likely to become successfully established in humans. The vector-borne diseases that can perpetuate themselves indefinitely in humans—the malarias, yellow fever, dengue—have come disproportionately from other primates. Other vertebrates have made an occasional contribution—sleeping sickness from ungulates, Chagas' disease from opossums, plague from rats—but these contributions are vastly less than would be expected if the contributions were proportional to the numbers of these other species. The greatest pool of closely related primates is in Africa, which also has its share of other potential sources, such as ungulates and rodents. We can therefore expect the dribble of germs that will enter and become established in humans to do so in sub-Saharan Africa more often than in other geographic areas.

Rodents do seem to be sending pathogens our way, but many of these rodent pathogens do not persist in humans. Two examples familiar to North Americans are the bacterium that causes Lyme disease and the hantaviruses that cause Four Corners disease. Rodents are probably common sources for such diseases because they live in such close association with us and in such great numbers. Occasionally they cause terrible damage, as occurred with the black plague, but even that disease fizzled out in the human population. That the threat from far-off places is only minor does not mean, however, that we have nothing to fear but fear itself.

CHAPTER SEVEN

Malignant Growths in Our Backyard

PARAMOUNT AMONG THE WEEDY PESTS IN OUR BACKYARD ARE the sexually transmitted pathogens, most of which crossed our borders decades or even centuries ago. The irony is that sexually transmitted pathogens are particularly dangerous even though their effects tend to be relatively mild on an average day during the course of an infection. The pathogens have to be fairly mild if they are to outlast the longevity of exclusive sexual partnerships. But for us the problem is not so much the threat on an average day as it is the threat over a lifetime.

HIV-1 infections give a sense of this difference. An infected person lives on average about ten years before developing serious disease. Almost all untreated infections will eventually be lethal. The probability of dying during a year of infection is therefore a bit less than 5 percent. Influenza infections generally cause a bit less than one death for every thousand infections. Influenza lasts about a week, so the probability of dying during a year of infection is also a bit less than 5 percent. Most people find AIDS more frightening than the flu even though an average person is much more likely to get the flu.

What does the historical record tell us about the danger from sexually transmitted diseases? For one thing, it shows that we have been paying the price of sexually transmitted diseases without knowing it. Long before the human papillomavirus was shown to be the primary cause of cervical cancer, women were getting the virus and dying from cervical cancer. That's why women need to get their annual Pap smear or risk lethal consequences. Until the 1980s it was generally not appreciated that women who were suffering and dying from cervical cancer were the victims of a venereal disease epidemic. Men need to be concerned too, if only out of self-interest—papillomaviruses are leading suspects in investigations of penile cancer.

But the shocker is that human sexual behavior appears to be changing disease organisms themselves, transforming them from harmful to mild or from mild to harmful depending on the details of our sexual behavior. Human papillomaviruses, for instance, can ratchet up their harmfulness within a year or so in response to greater opportunities for sexual transmission. They can change from an innocuous, albeit embarrassing, cause of warts into a killer.

The most cancer-causing of the human papillomavirus types create havoc by using proteins called E6 and E7. These proteins neutralize our bodies' safeguards against cancer. E6 binds to a protein called p53, which guards against cancer by arresting cell division. If E6 fails to take out p53, however, the human papillomavirus can still neutralize p53's protective effects using E7, which binds to a second anticancer sentry stimulated by p53. E6 also interferes with the cell's self-destruct mechanism, which can protect the rest of the body from pathogens in an infected cell. By these means the papillomavirus is able to keep the cells working in the interest of the virus rather than in the interest of the person.

The lethality of cancer is as devastating to the virus as it is to the person, but the path leading to cancer is a clever solution to a difficult problem. Being sexually transmitted, the human papillomavirus has to deal with the fundamental constraint of sexual transmission: it has to be sufficiently persistent in sufficient numbers to capitalize on future opportunities for transmission. It needs to hide out from a primed and vigilant immune system. The clever solution of the papillomavirus is to keep the cell multiplying and then multiply right along with it. The more the cell divides, the more the virus can replicate itself with little exposure to the surveillance of the immune system.

But how much of the racketeering is too much even for the virus? That depends on the opportunities for transmission. Little exposure to the immune system does not mean no exposure because even a small amount of exposure can make the virus vulnerable to the immune system's sophisticated virus detection technology. When a virus is clandestinely active within a cell, the cell will post fragments of the virus on its outside walls. It's like the warning sign posted on the walls of a quarantined house to notify the rest of the community. But the actions taken inside our bodies are not tempered by a liberal mind-set—extremism in defense of the body is no vice. When members of the immune system see the sign, they do not just stay away from the cell; they blow it up. The notice does not read KEEP OUT! but rather KILL ME! By posting this notice of infection, the cell is committing suicide to protect the other cells in the body from infection. For the most cancer-causing of the papillomavirus types, parts of the viruses' E6 and E7 proteins are posted on the outside of the cell to attract the friendly fire of the immune system. These extreme responses by the body help control the infection in most women before it develops into cervical cancer.

So even a papillomavirus that is clandestinely reproducing as the cells are reproducing may not be clandestine enough. Those viruses programmed for less of this manipulative reproduction may be less likely to attract the immunological attack. By keeping an extremely low profile, such viruses might increase the chances of their being around in the reproductive tract when a new sexual partner is selected. The suppressed activity of the virus translates into a lower probability of infection per instance of sexual contact, but a sexually active person who is not changing sex partners frequently will tend to have more sex with each partner; the sex is just spread out over a long period of time. So the virus's low profile may still yield a high probability of infection per partner—if the partner is not infected today, then maybe next week, next month, or next year. That strategy is better for us because a virus that keeps a very low profile is a virus that is very unlikely to cause cancer.

On the other hand, if people change partners frequently, then a virus with such self-restraint pays the price of missing transmission opportunities. In a promiscuous society, the papillomaviruses that trigger more rapid reproduction would tend to spread more effectively to new hosts. The more rapid reproduction may make the viruses less cryptic and hence more vulnerable to the immunological defenses of the current host, and more likely to cause a cancerous death or to fall victim to surgical excision, but the reduced tenancy in any single person is offset by the increase in transmission to new people. When opportunities for sexual transmission are high, the virus reaps a high return on its high rate of replication. The dividend is increased representation of the nasty types of papillomavirus in the overall papillomavirus population. When opportunities for sexual transmission abound, human papillomaviruses should evolve toward high virulence.

This is not the kind of statement that can be tested with controlled experimentation. It would be a no-brainer for an NIH ethics committee. But human behavior, being what it is, gives us "natural experiments" that would be unethical had they been planned and executed. One of the most horrific of these situations occurred during the war in what was until recently Yugoslavia. Opportunities for venereal transmission typically increase during wartime, but in that war the opportunities were even greater because rape was used systematically as a weapon. If human papillomaviruses evolve increased harmfulness when the opportunities for transmission increase, they should have done so over the course of the war.

Evidence indicates that they did. Just before the war researchers found that among sexually active women in the region, the most lethal papillomaviruses were less common than the milder types. But by the end of the war these lethal types outnumbered the more benign types by three to one. The vicious viruses increased tenfold over the two years of the war; the benign types remained close to their prewar level. By the third year after the war, the lethal types had begun to recede.

Wartime conditions are extreme. Is there any evidence that the harmfulness of human papillomaviruses depends on sexual behavior in peacetime? Women who have more sexual partners do tend to have more papillomavirus infections. This trend holds from Sweden to Brazil to Colorado. But the risk is not the same for all types of papillomaviruses. The details are still unclear because sufficiently sensitive techniques have been applied only during the past few years, but the current evidence indicates that having more sexual partners increases disproportionately the risk of acquiring the most dangerous papillomaviruses. In Brazil during the mid-1990s, for example, the risks of acquiring the harmful and benign viruses were similar among women who had no more than five sexual partners during their lifetime. Women who had more than five but not more than ten were almost twice as likely to harbor the most dangerous types. Of the women who had more than ten partners, one third were infected with the most dangerous types, three times as many as among the women who had no more than five partners during their lifetime.

The broader implication of these trends is clear: if people start having sex with more people or with less protection, venereal pathogens will not only spread but will evolve to become more harmful. This point makes clear why the threat of disease emergence is largely a homegrown problem. The raw materials for harmful and mild infections are already globally distributed; they are continually being reseeded by rapid air travel as well as by biological mutations and recombinations, but these processes only provide the raw material. Whether we have a grave problem will depend on whether our local soil—our own behavior—favors the harmful forms over the mild forms.

We noticed the AIDS pandemic while it was spreading, but we missed the initial spread of all the other known and unknown infectious causes of cancer and infertility. For most of these causes we missed the invasion because it happened long ago. We might have missed the initial spread of the AIDS pandemic had it not occurred early and disproportionately among homosexual men because they tend to be seen as a distinct group in our society and were therefore readily categorized as a risk group. If attention had focused on the syndrome in 1991 rather than 1981, we would have seen HIV in almost every country on earth. The situation would have been like the current situation with the cancer-causing papillomaviruses, or with hepatitis viruses, which cause levels of death and suffering that are comparable to those now caused by HIV.

Some might say that the deaths from AIDS surely would have led people to recognize promptly the infectious nature of AIDS. But countless thousands of people were dying of liver cancer, cervical cancer, and adult T-cell leukemia long before these diseases were recognized as being caused by infection. And countless thousands have died and are now dying of other cancers that will probably turn out to be caused by infection. Even one of the main indicators of AIDS, a cancer called Kaposi's sarcoma, had been recognized in Mediterranean populations a century before it was recognized as an infectious disease. The presence of Kaposi's sarcoma in gay men with AIDS, and its relative absence in other risk groups, alerted researchers to the probable role of infection in its causation. Still, it took a decade of searching for an agent of Kaposi's sarcoma before researchers finally zeroed in on human herpesvirus 8, newly recognized but long ago globally distributed.

THE BURDEN OF PROOF

In 1874, eight years before Robert Koch presented his discovery of the bacterial agents that cause tuberculosis, a lesser-known microbe hunter,

Arthur Boettcher, published a paper on a small, curved bacterium that he found repeatedly in ulcers of the stomach. Over the next half century, several other scientists confirmed Boettcher's finding. Some also extended the research by experimentally transmitting the bacterium in lab animals. By the late 1940s peptic ulcers were being successfully treated with antibiotics in New York City hospitals. Paul Fremont-Smith was a young intern in 1948 at Manhattan's New York Hospital. He remembers the orders he was given there by his no-nonsense supervisor, Connie Guion. She told him, "The people over at Mount Sinai have found that Aureomycin [a trade name for chlortetracycline] is effective against peptic ulcers. Use it. It works." He did, and it worked. Then, around 1950, discussions of infectious causation of ulcers disappeared from the literature and from the treatment regimen. The medical texts from 1950 through the early 1990s attributed peptic ulcers to gastric acidity, stress, smoking, alcohol consumption, and genetic predispositions—everything but infection. Generally there was not even a reference to the possibility of infectious causation.

Around 1980, after three decades of medical impotence against ulcers, researchers in Perth, Australia, were back on the infection trail. The driving force was a young internist named Barry Marshall. He was not guided by the long history of research on the subject, which began with Boettcher and ended mysteriously around 1950. In fact, Marshall was unaware of this history—medical education in the 1970s, like medical education in the year 2000, did not waste time making medical students learn about the mistaken theories of the past.

Like Boettcher over a century before, Australian pathologists during the late 1970s noticed the curved bacteria in tissue samples from ulcer patients. Like Connie Guion over three decades before, Marshall in 1981 saw his patients with ulcers and gastritis improve after tetracycline treatment. He got the attention of microbiologists and histologists in his hospital who were needed to resolve the matter. They found the curved bacilli in all their duodenal ulcer patients and in 80 percent of their peptic ulcer patients. In April 1982 they cultured the bacterium that caused the trouble, now known as *Helicobacter pylori.*

Infectious causation of peptic and duodenal ulcers is now widely accepted by medical authorities, though this acceptance has grown only gradually since the mid-1980s. Marshall helped the process along by intentionally drinking an infective dose of *H. pylori*, getting gastritis, and then curing it with an antibiotic. Still, it was only in the mid-1990s that the medical establishment finally generally accepted the idea that peptic and duodenal ulcers are infectious diseases.

It is not clear why almost all medical experts ignored for four decades the evidence that ulcers could be caused by infection. Several attributes of ulcers made infectious causation inconspicuous: a loose correlation between infection and ulcers, the internal site of infection, and variable delays between the onset of infection and the onset of overt disease. Partly as a result of these attributes, evidence of infectious causation could be overlooked or dismissed. *H. pylori* is so cryptic in its disease causation that its transmission mode is still unclear today.

Another reason for the delay in solving the ulcer puzzle stems more from psychology than medicine. Some researchers reported in the late 1940s that they could not find the curved bacteria in fresh specimens; they attributed the associations that were found by others to bacteria that were growing in the ulcers after death. Perhaps this kind of negative evidence was all that was needed in a time when medical thinking was influenced by Freudianism, which emphasized the power of mind to influence disease and the fragility of that powerful mind in response to stressful influences of the social environment. If the mind, altered by the stress of twentieth-century life, could make a person mentally self-destruct, certainly it could cause a little old ulcer.

Besides, infectious causation was beginning to be old hat; genetics had become the biological answer to disease, especially after the structure of DNA was discovered in 1953. This discovery would eventually allow an understanding of what went wrong brick by brick, from the nucleic acid bricks that were used to build the DNA and RNA, to the amino acid bricks that were used to build the proteins the DNA and RNA encoded. Using this metaphor, genetically minded medical researchers thought that they could understand why biological buildings had problems and eventually solve those problems. The history of the subsequent half century shows that this general approach was often correct, though sometimes misleading. Genes that cause human disease were found, but often those genes were not human; viral genes were found inside cells, sometimes even integrated into human DNA. The structures these genes encoded were not buildings at all, but tanks, missiles, and sinister devices of sabotage and manipulation, such as the E6 and E7 proteins of human papillomaviruses.

Still another reason for the slow acceptance of infectious causation of ulcers stems from the standards of evidence used in medical science. Simply, the standards of evidence were too high. When standards of evidence are set too high, scientific rigor declines. It is easy to recognize this problem when the standards are set so high that they can never be met. If, for example, scientists demanded experimental demonstration of humans evolving from ancestral apes before accepting the hypothesis that humans evolved from apes, the hypothesis could never be accepted. This too-high standard of evidence would have sapped the rigor of scientific inquiry into human origins. The same logic applies when the standards are extremely difficult but not impossible to reach, though the trade-offs between standards of evidence and scientific rigor is more difficult to assess. The required shift in standards seems especially difficult for researchers to accept when the standard has been reached for somewhat similar hypotheses. This is the state of affairs in studies of infectious causation.

Standards for identifying infectious causation had been established during the 1880s when infectious causes of acute diseases were being rigorously identified at a remarkable pace. But the early success of these standards led health scientists to overlook the importance of a simple fact: diseases caused by infection will vary greatly in the degree to which their infectious causation can be demonstrated by any particular set of standards. Applying one set of standards will allow early recognition of diseases that are easily recognized by the standards, and leave in the wake of that recognition those diseases that cannot be so identified. Throughout most of the twentieth century the experts maintained a stalwart grip on has-been standards that had largely outlived their usefulness.

WHERE TO LOOK

What kinds of pathogens cause the damaging chronic diseases we don't know about? What should we look for? Persistent infections are more likely to cause chronic diseases than short-lived infections because the more time a pathogen has to gum up the body's machinery, the more likely it is that it will. Internal pathogens are more likely to cause problems than external ones because internal tissues are more delicate and more central to life processes; moreover, internal pathogens must cope with the immune system both to survive and to get out—and this ups the ante. Their persistence in the host provokes immunological destruction of infected cells and inevitable casualties from friendly fire. The tricks that allow a pathogen to avoid immunological destruction can push our immunological systems to the point of self-destruction.

These criteria fit those pathogens that have been traditionally classified as sexually transmitted. They also fit pathogens that are transmitted by less hard-core sexual contact, such as the Epstein-Barr virus, which causes infectious mononucleosis and is transmitted by intimate kissing. Sexually transmitted diseases make up only a small fraction of human diseases, but they are major players in the known chronic diseases caused by infection. Syphilis, infertility due to *Chlamydia* and gonorrhea, arthritis and ectopic pregnancy caused by *Chlamydia*, and cervical cancer are just a few of the mentally or physically debilitating chronic diseases caused by sexually transmitted pathogens. Sexually transmitted pathogens are likely to cause a disproportionately large number of the infectious chronic diseases yet to be discovered. We already know of some suspects. Human parvovirus B19, for example, probably contributes to the crises of sickle-cell anemia and to chronic diseases such as arthritis and multiple sclerosis. Some recent evidence indicates that it may be at least partly a sexually transmitted pathogen. In 1999 researchers in the Italian Red Cross reported a comparison of parvovirus B19 infection rates among patients in an STD clinic with rates among healthy blood donors who were used as controls. Almost 40 percent of the clinic patients were infected with B19, compared with only 10 percent of the blood donors.

The other infectious causes of chronic diseases will probably be an eclectic collection much like those that have already been discovered—a grab bag of pathogens. Some of these pathogens may rely on their persistence for transmission, like *H. pylori* and the papillomaviruses. Others may just by chance have characteristics that allow them to persist in the body in places unrelated to transmission; this category is illustrated by chronic ear infections that are caused by pathogens of the respiratory tract, such as *Haemophilus influenzae* and *Streptococcus pneumoniae.* Still other pathogens may cause chronic disease because of long-term disturbances of the body's machinery that persist even after the pathogen is gone. Some autoimmune responses might fall into this category, such as chronic heart disease caused by streptococcal infections.

ACROSS THE SPECIES DIVIDE

Most of the attention-grabbing emerging diseases are caused by pathogens transmitted to humans from other species, the so-called zoonoses. Reemerging diseases such as cholera, influenza, and dengue are included in discussions, but it is the more exotic zoonotic diseases, such as Ebola, that seem most intriguing and frightening—probably because they raise the possibility of previously unknown horrors sweeping through the earth's human populations, much as AIDS has done. In contrast, most of the chronic diseases with recently recognized infectious causes are caused by pathogens that have been transmitted from person to person for centuries or even millennia.

Most, but not all. Lyme disease, which causes most of its damage while in its chronic state, is a zoonosis. In humans it causes only dead-end infections; it enters humans from field mice through the bite of a tick, causes arthritis and a variety of other chronic problems, but never moves on from humans to infect other humans. The presence of Lyme disease is therefore restricted because of the restricted distribution of the rodent host.

Zoonotic diseases tend to have geographic distributions that are restricted to the geographic distributions of their reservoir hosts, and the chronic diseases for which infectious causes are suspected tend to be widespread. One might therefore conclude that important chronic diseases are unlikely to be caused by zoonotic pathogens and restrict the search accordingly. This restriction could be a mistake because many animals that might harbor dangerous diseases have planetary distributions mirroring the distribution of humans. The hosts are our domesticated and domiciliary animals, such as dogs, cats, horses, cows, sheep, chickens, and the party crashers who live with us without an invitation—rats and mice.

Epidemiologists who study the evolutionary origin of infectious diseases identify such animals as the ultimate source of full-fledged human pathogens, such as the agents of tuberculosis and measles. Such zoonotic sources are also widely implicated in dead-end, acute infections—for example, toxoplasmosis from cats, glanders from horses, and rat-bite fever. There are now indications that this category may include some of the most problematic chronic diseases as well.

Consider breast cancer. Infectious causation has been virtually ignored by most experts on breast cancer despite the infectious causation of mammary tumors that is the rule rather than the exception among other mammals. Two research groups have recently found evidence of retroviruses in breast tumors but not in surrounding healthy tissue. One of the research groups, based at the Mount Sinai School of Medicine in New York, compared the viral sequences with sequences of previously identified viruses and found them to be indistinguishable from those of a virus that causes tumors in mice, the mouse mammary tumor virus. More than one third of the human breast cancers sampled by the group were positive for the virus, whereas less than 2 percent of the normal breast tissue had evidence of viral genes. The geographic distribution of breast cancer mirrors the geographic distribution of the virus's primary host—the house mouse *Mus domesticus*—suggesting that breast cancer may be in part a disease acquired from mice. The match is not perfect, but a perfect match is not necessarily expected for diseases that may have a variety of causes. In the mice, transmission occurs both by sex and through mother's milk, as is true for the distantly related human retrovirus HTLV-I. The other leading suspect is the Epstein-Barr virus, which has also been found disproportionately in human breast cancer.

Although the known genetic causes of breast cancer have received much more attention, they account for only about 5 percent of breast cancer risk. And even this genetic risk is not necessarily independent of infectious causation. People who are genetically predisposed to get a disease may be genetically predisposed to get a particular infection that causes the disease rather than getting the disease directly from faulty genes. Studies of identical twins show just how much of a genetic basis there could be. In spite of all the attention to genes for breast cancer, the concordance between identical twins is low. This fact is well illustrated by a study of women in Wales and England conducted by epidemiologists at the London School of Hygiene and Tropical Medicine. Among women twenty to thirty years old with breast cancer, the probability that an identical twin also had breast cancer was 25 percent; the concordance was below 10 percent for identical twins older than thirty-five. Interestingly, the concordance was not lower among fraternal twins, as typically occurs when a phenomenon is a direct result of host genes. A woman's genes are of some importance, but other things are much more important. The balance of evidence implicates infectious things.

CHAPTER EIGHT

Our Vulnerable Hearts and Minds

IN 1965, ABOUT FIFTEEN YEARS BEFORE MARSHALL AND HIS Australian colleagues were piecing together the role of *H. pylori* in ulcers, a much smaller bacterium was isolated from the eye of a child in Taiwan. This bacterium would incite investigations into the infectious causation of chronic diseases far more deadly than peptic ulcers. In the 1980s, while Marshall and his colleagues were wondering why medicine was not paying attention to their evidence for infectious causation of ulcers, this other line of inquiry traveled a circuitous route, passing through critical junctures in Seattle, South Africa, and Finland. This route, which is continuing in unforeseeable directions, may lead to one of the greatest medical advancements since the development of antibiotics in the middle of the twentieth century.

The research in Taiwan was led by Thomas Grayston, a medical microbiologist with a better-make-sense-if-you-want-my-attention attitude. He gets to the heart of a matter quickly. If he disagrees, he will say so, without wasting time and without doing much to soften the blow. Yet in the tradition of others who have guided the health sciences through new frontiers, he balances his hard-nosed insistence on clarity of expression and solid evidence with an appreciation for the value of new perspectives and fresh thinking.

Grayston's group originally thought that the organism in the eye of the child was *Chlamydia trachomatis,* the same *Chlamydia* featured on pamphlets in STD clinics. *Chlamydia trachomatis* had been seen in tissue samples since the first decade of the twentieth century, but a much longer history in humans is evidenced by some ancient medical paraphernalia: special spoons made about two thousand years ago in the Middle East. The spoons were made for people who had trachoma, which is a disease characterized by a puffy swelling of the eyelids that rotates the eyelashes, pushing them onto the eyeball. With every blink the lashes brush the eye. The continual brushing eventually abrades the eye, causing scarring, invasion of the cornea by blood vessels, and often blindness. The spoons were set concave side down on the eyelids to force the eyelashes away from the eyeball. Children infect their eyes when they rub them with contaminated fingers. The infection is apparently maintained by eye-finger-eye transmission, perhaps supplemented with transmission from fingers or fabrics inadvertently contaminated with genital secretions.

Grayston's group originally thought that their organism was *C. trachomatis* because trachoma was widespread in poor countries. Worldwide, about 400 million people had the disease, and the infection was far more common. But their isolate did not grow well in cell cultures that supported a florid growth of *C. trachomatis*. In 1986 they published their findings in the *New England Journal of Medicine*, proposing that they had discovered a new variant of the bird parasite *Chlamydia psittaci*.

Having taken the problem back to their laboratory at the University of Washington in Seattle, Grayston's group was attempting to figure out this problem when they were visited by a young Finnish researcher named Pekka Saikku. Together they began testing different populations of people for antibodies to the unusual *C. psittaci*–like organism. They found antibodies in Seattle, in Helsinki, and everywhere else they tested. By the time people reached middle age, over half of them showed signs of infection. That seemed strange because *C. psittaci* was thought to be transmitted only from birds to humans, and not from person to person. In the past it had been found in people who owned pet birds and in poultry farmers, but rarely in the general public. This new variant was far more widespread than the standard version of *C. psittaci*. Moreover, the pathogen was not just a benign freeloader. It was responsible for about one of every five cases of pneumonia.

The Grayston lab found that the DNA of the *C. psittaci*–like bacterium was just as distinct from the known strains of *C. psittaci* as it was from the known strains of *C. trachomatis*. In 1989 the Seattle researchers named the bacterium that they had first isolated from the Taiwanese child's eye *Chlamydia pneumoniae*. A combination of epidemiological and microbiological study revealed that *C. pneumoniae*, like *C. trachomatis*, used humans as its primary host. But unlike *C. trachomatis*, it was a respiratory tract pathogen; it infected the lungs and was transmitted to the lungs of other people via droplets expelled into the air by coughing.

As Grayston's group was moving toward this conclusion, Pekka Saikku and Maia Leinonen were taking their analysis of Finnish patients in a direction that still has experts in the health sciences vehemently arguing. With one of the highest rates of heart disease in the world, Finland has many labs working cooperatively on the problem. To assess the presence of infections after heart attacks, a large sample of blood serums was collected by researchers at Helsinki University Central Hospital and distributed to several labs in the area. Working at the National Public Health Institute of Helsinki, Maia Leinonen analyzed the samples for *Chlamydia*. She found that 70 percent of the samples from heart attack patients had antibodies against a compound common to *Chlamydia*. This percentage was significantly higher than the percentage in the control serums from people who had not had heart attacks. Only one *Chlamydia* species could produce rates so high: *C. pneumoniae*. There was no evidence, however, of recent infection, and the heart attack rates in Finland did not follow the six-to-seven-year cycles of acute *C. pneumoniae* outbreaks. This evidence led Leinonen and Saikku to hypothesize that heart attacks and the underlying atherosclerosis that causes them were caused by chronic *C. pneumoniae* infections.

Responses ranged from fascination to ridicule. Strong responses are appropriate to any suggested explanation of a disease that eventually kills about half of all the people who live in wealthy countries, usually by leading to a heart attack or a stroke. Atherosclerosis is that lethal. The jarring implication of Saikku and Leinonen's hypothesis is that humans have never escaped plagues of infectious disease. Even the bubonic plague did not claim such a toll. If their hypothesis is correct, the primary differences between the notorious plagues of the past and this plague of the present are the slow, insidious pace of cardiovascular disease, and the mature age at which it strikes—in short, the plague's timing.

Unscientific dismissiveness and ridicule in the medical world has gradually given way, as the evidence has mounted, to scientific criticisms based on alternative hypotheses. It was not just the exciting promise of Saikku and Leinonen's hypothesis that attracted attention. It was also the historic stature of the problem of cardiovascular disease and the many decades and billions of dollars that researchers have spent studying it from different perspectives—particularly diet and genetics. Researchers have taken sides on the issue and are often as entrenched as the soldiers on the front lines in France in 1916.

One criticism of the infection hypothesis concerns the old problem of causation and association. Proof of an association does not represent proof of causation. A hypothetical causal mechanism is the first step in addressing this criticism. One possibility is that the organism is actually at the site of atherosclerotic damage, in the artery linings, either damaging the arteries directly or triggering an immunological response that does the damage. If either possibility were true, one should be able to see the *Chlamydia* at the site with microscopy. In 1992 the South African pathologist Alan Shor was studying sections of atherosclerotic plaques from arteries excised during bypass operations. There in the pictures from his electron microscope he saw bacteria with the telltale pear shape of *C. pneumoniae.* He sent some of his samples to the Seattle lab, which used an array of techniques to confirm definitively the presence of the organism.

The causation-versus-association problem does not go away so easily, however. Perhaps *C. pneumoniae* is simply an innocent bystander. After all, the bacteria are inside white blood cells, and the white blood cells are in places where there is damage. Perhaps something else is causing the damage, and *C. pneumoniae* is just there because that is where the action is.

One way of addressing this issue is antibiotic treatment. Models exist for using this avenue toward resolution. It was antibiotic treatment, for example, that resolved the half century of controversy over infectious causation of rheumatic fever: when streptococcal infections were cured with antibiotics, rheumatic fever did not occur. Interpretations of antibiotic treatment are not as straightforward for most chronic infections, however. Treatment of bypass patients with antibiotics, for example, improves their recovery. But is the antibiotic treatment controlling *C. pneumoniae* or some other bacterium hindering recovery? Past treatment with tetracycline has been associated with lower rates of first-time heart attacks in one study but not in another. Discrepancies among such studies is not surprising, considering that many variables can vary from place to place. Still, even if an antibiotic reduces the rate of heart attacks, one can always argue that it is having other effects that may be ameliorating the damage. Researchers have noted, for example, that any improvement associated with tetracycline could be attributed to its anti-inflammatory effects rather than to its antimicrobial effects.

The role of inflammation may prove to be the most difficult part of the heart disease puzzle. Many researchers are considering the hypothesis that atherosclerosis is an autoimmune disease in which the damage is caused by inflammation. This begs an important question: What causes the inflammation? From an evolutionary point of view, it does not make sense that the immune system would start malfunctioning on its own in a high proportion of people. For the few common diseases in which autoimmunity is well understood, an infectious agent is the cause of the trouble. Researchers who favor the inflammatory self-destruction hypothesis tend to ignore this point.

And sometimes "ignore" is putting it too mildly. When the journalist Philip Ross was preparing a story on atherosclerosis for *Forbes* magazine, a leading researcher told him that infectious causation of atherosclerosis was "bullshit." (This researcher's published treatments of the hypothesis tend to be more tactful.) The researcher favored the noninfectious inflammation hypothesis, with a focus on C-reactive protein, which is a marker of inflammation and helps control bacterial infections. He mentioned that *C. pneumoniae* infection was present in the atherosclerotic plaques because white blood cells are present and *C. pneumoniae* infects white blood cells. That bystander hypothesis deserves testing, but it is not evidence against infectious causation of atherosclerosis by *C. pneumoniae.* The researcher conveniently overlooked the pressing question: why should the body's immune system go beserk, causing inflammatory damage?

The standard answers—"The body isn't perfect" and "The immune system is so complex, it's bound to go haywire" —just won't do. There are many people who have immune systems that do not go haywire. Over time the instructions for these reliable immune systems should have displaced the people whose immune systems did go haywire. If a more complex immune system often goes haywire, then how could the more complex system evolve from a simpler system? This imperfection-of-complexity argument might be rescued if the more complex system is necessary to better cope with other pathogens; in this case the compromises imposed on an immune system by these other pathogens could make the system vulnerable to going haywire through an autoimmune mechanism. Infection would still be involved, but less directly. Though this explanation could be true, it seems like jumping through hoops to avoid the simpler, more parsimonious, explanation—namely, that infection is causing the damage directly, as well as through the inflammation it triggers. Another problem with the argument based on immunological compromise is that people who suffer from atherosclerosis do not appear to be protected against other diseases. In fact, they seem to be more prone to other diseases, such as Alzheimer's, on the basis of their genetic makeup. The low concordance among identical twins for common autoimmune diseases indicates that something else is important, either infection or some noninfectious environmental influence. When analyzed deep down to its premises, inflammation by itself lacks a conceptual foundation. Infectious causation provides the inflammation hypothesis with such a foundation because the infectious causation hypothesis does not specify whether the damage is caused directly by the organism or indirectly through the organism's stimulation of an inflammatory response.

Indeed, inflammatory processes are often a clue to infectious causation. When did researchers first start studying the inflammatory process in atherosclerosis? In the 1820s. When did researchers first propose infectious causation of atherosclerosis? In the 1870s, almost as soon as infectious agents were discovered. When did researchers first find evidence that *Chlamydia* was involved in arterial disease? In the 1940s. When was the idea of infectious causation of atherosclerosis dropped? In the 1950s. This should be sounding familiar. The ball was dropped for atherosclerosis much as it was dropped for peptic ulcers, and at about the same time.

The infectious causation ball was picked up again for atherosclerosis about ten years after it was picked up for peptic ulcers. One might expect then that if the processes of debate and experimentation are similar, infectious causation of atherosclerosis could soon be generally accepted, perhaps within the next few years. But the proposal of infectious causation of atherosclerosis probably will not be accepted so soon for several reasons. If *C. pneumoniae* does cause atherosclerosis, the demonstration of this role with available tools will be more difficult than the demonstration of a role for *H. pylori* in peptic ulcers. The circuitous route from a child in Taiwan to the present has now generated about forty studies that have tested, in one way or another, the hypothesis that *C. pneumoniae* is linked to atherosclerosis. About three quarters of them have confirmed the link. This track record might not seem particularly impressive, but the tools that are being used are not well developed for the task at hand.

The detection of the specific antibodies that the immune system manufactures in response to a given pathogen has been a reliable indicator of infectious causation of acute infectious diseases over the past century. Using these antibodies to detect chronic *C. pneumoniae* infection is only marginally effective, however, because the *C. pneumoniae* in chronic infections may often be sheltered from the immune system. *C. pneumoniae* lives within our cells, is often covered by plaque, may often be inactive, and may be triggering antibodies largely as a result of lung infections.

Similar difficulties apply to identifying of *C. pneumoniae* by growing it in nutrients to obtain sufficient numbers of bacteria for tests. This standard method for identifying bacteria is not particularly useful for *C. pneumoniae* because techniques for isolation of *C. pneumoniae* from atherosclerotic plaque are not available for living tissue and are difficult even when fresh specimens have been obtained from surgery. If the organism cannot be readily isolated and cultured, associations between the presence of the organism and the presence of disease are difficult to establish.

These barriers are special cases of a general phenomenon that can be expected when evaluating infectious causation of chronic diseases: the adequacy of a given tool for ascribing infectious causation should decrease as it is successfully applied. This decline in effectiveness occurs because examples of infectious causation that can be readily identified with such tools become so identified, leaving in their wake those infectious causes that are less detectable. Koch's postulates were good for identifying the agents of acute infection. They are often more difficult to fulfill in cases of chronic infection, especially when the suspected agent is in tissues that are difficult to reach. The success of tools drives the tools to the edge of their utility. As this process occurs, assessments should take the whole research picture into account rather than focusing exclusively on a few studies, until better tools can be developed.

THE PLAGUE OF THE HEART

Better tools are needed because atherosclerosis is an insidious, multifarious disease. It is characterized by fatty deposits that start developing as streaks just inside the lining of arteries. Recent studies indicate that some streaking can be found even in teenagers. Streaks are acquired by an increasing proportion of the population from then on. About one quarter of people in their late twenties have acquired them, and most have acquired them by the time they are in their sixties. As a person ages, the streaks grow into fat deposits. The growing blob inside the arterial wall pushes the lining of the artery inward into the lumen of the vessel, constricting blood flow and weakening the artery lining. If the lining breaks, the blob can proceed into the lumen of the artery, clogging it there, or it can be carried to other parts of the body. If the clogging occurs in the coronary artery, which supplies the heart with blood, the result can be a heart attack (myocardial infarction)—the heart cells that are supplied by that vessel cannot receive oxygen and die. If the clogging occurs in a blood vessel in the brain—a stroke—the brain tissue supplied by the artery similarly dies.

If one asks the experts what causes atherosclerosis, the answer depends on the expert. Partly to develop a consensus on the matter, modern medicine has shied away from statements of causation. Although researchers often cannot agree on causation, they can come close to a consensus on a less informative concept: risk factors. Studies can show, often fairly unambiguously, that certain aspects of people's lifestyle and biology are statistically associated with increased risk of atherosclerosis. When such an association is unlikely to occur by chance, the characteristic under consideration is accepted as a risk factor for the disease. This approach has identified an array of risk factors for heart attacks and stroke, including smoking, a high-fat diet, high cholesterol levels, high iron levels, high blood pressure, markers of inflammation, and genes.

But the weakness of the approach lies in exactly the same characteristic that allows it to be accepted by the spectrum of experts—it skirts the issue of causation. This is a weakness because knowing causation is far more useful than knowing association. All risk factors are not created equal. If one risk factor is the primary cause of a disease, then eliminating that risk factor should prevent the disease. None of the risk factors for atherosclerosis listed above appear to be a primary risk factor. For each of these risk factors, many people can be found who do not have it but still come down with atherosclerosis. The recognition of this fact, together with the protection from criticism that is gained from speaking of risk factors rather than causation, has led to a strange state of affairs in the study of atherosclerosis, as well as the study of other chronic diseases of unknown cause. Many researchers have begun to think that such diseases originate as some sort of a composite sum of the various risk factors, with each one contributing according to its relative risk.

This line of reasoning runs contrary to the track record of great achievements in medicine, which have all resulted from principles of primary causation. This conclusion holds for the great theoretical frameworks, such as the germ theory of disease, as well as for the great practical interventions, such as surgery, antimicrobial drugs, vaccines, improved nutrition, and hygiene. With this track record of achievement, it seems astonishingly unwise to retreat from attempts to explain disease by principles of primary causation just because these principles have not been identified and may be controversial.

This admonition is particularly relevant to atherosclerosis. If all the noninfectious risk factors are combined, they explain only about half the risk of acquiring atherosclerosis. In other words, about half of the people with atherosclerosis acquire it even though they do not have elevated risk factors for the disease. Something big is missing from the picture of risk factors.

The test of primary causation is to see whether or not the disease is absent when the proposed primary causes are absent. This test is easiest to visualize with infectious diseases. In the early 1880s, when medical experts were just beginning to consider the bacterium *Mycobacterium tuberculosis* as the primary cause of pulmonary tuberculosis (then called phthisis), they had assembled an impressive array of risk factors. One of the standard texts of the day, Quainn's *A Dictionary of Medicine*, listed family predisposition, acute febrile diseases, syphilis, mental depression, bad ventilation, climatic influences, dampness of the soil, inoculation, and debilitating conditions. The last category was a big one; it included "miscarriages, unfavourable confinements, over-lactation, insufficient food, and alcoholism." Altogether, fourteen subheadings of causation were bulleted. One of these sections was headed "Infection." Although Robert Koch had identified *M. tuberculosis* as the cause of tuberculosis two years before, this two-paragraph section begins, "The idea of infection being a cause of phthisis still prevails in the South of Europe, and has lately been revived by Dr. Budd in England." Hooray for Dr. Budd! The section goes on to describe anecdotal evidence of infectious transmission. The section on infection amounted to about one eighth of the discussion of causal factors. *A Dictionary of Medicine*, like other texts of the time, failed to make a critical distinction. Primary causes can be separated from risk factors. *M. tuberculosis* is a primary cause of tuberculosis—if *M. tuberculosis* is not present, pulmonary tuberculosis does not occur no matter how many of the other thirteen risk factors are present.

This significance of this illustration should not be dismissed as a symptom of the backwardness of medicine in the 1880s, unless one is ready to dismiss similarly the backwardness of medicine a century later. In the fifteenth edition of the *Cecil Textbook of Medicine*, published in 1979, peptic ulcers were attributed to a similar collection of risk factors: gastric acidity, stress, smoking, alcohol consumption, and genetic predisposition. A decade later, and several years after Marshall and Warren reintroduced the idea of infectious causation, little had changed. A paper in the leading journal *Gastroenterology* concluded, "Our study supports the concept that several interacting factors (psychologic, behavioral, and genetic/physiologic) are likely involved in peptic ulcer disease. Emotional stress may predispose to ulcers by producing gastric hypersecretion, as manifested by hyperpepsinogenemia." Infection was not mentioned.

Recent texts treat atherosclerosis in the same way. The current resistance to a fair consideration of infectious causation of atherosclerosis thus seems like déjà vu to some observers. But does infectious causation of atherosclerosis really have the potential to bring together in a coherent picture the morass of risk factors: smoking, high-fat diet, high cholesterol levels, high iron levels, exercise, high blood pressure, markers of inflammation, and host genes? Yes. Here are some examples.

The major accomplishment of the research into the genetic causes of atherosclerosis concerns a gene called apolipoprotein E (ApoE), which codes for alternative forms of a protein that is involved in fat transport. ApoE has three common alternative forms, known as alleles; they are called ϵ2, ϵ3, and ϵ4. The ϵ4 allele has been identified as a risk factor for atherosclerosis and stroke. Researchers whose attention is fixed on host genes tend to think of ϵ4 as an inferior allele that causes this damage because it is not as good as the other alternative alleles. But there is a problem with this explanation. If it were the whole story, ϵ4 would not be around at anywhere near its present frequency, which amounts to about 20 percent of the ApoE alleles. Even though the negative effects of atherosclerosis and stroke tend to occur in old age, the negative effects of ϵ4 are significantly negative to cause it to vanish from the gene pool, or to keep it from arising in the first place. The situation must be more complicated.

Some clever research on arthritis patients provides a clue. The fluid inside the knee holds valuable information. By analyzing the genetic material in the fluid, molecular biologists can discover not only the ApoE instructions a person is carrying but also pathogens that may be residing in the knee, pathogens such as *C. trachomatis* and *C. pneumoniae. C. trachomatis* is known to cause some forms of arthritis, and *C. pneumoniae* is suspected. In this study, people who were not infected with either of these species of *Chlamydia* had the ϵ4 allele at the normal frequency. People who were infected with *C. trachomatis* also had ϵ4 at the normal frequency. But people who were infected with *C. pneumoniae* were three times more likely to be carrying the ϵ4 allele.

This finding casts new light on the role of ϵ4 in atherosclerosis: rather than causing damage because it is simply an inferior allele, it may cause damage indirectly by increasing a person's vulnerability to *C. pneumoniae.* This answer resolves the persistence paradox (if a gene is bad for us, it shouldn't persist in our species). Because infectious agents are involved in coevolutionary battles with their hosts, the "good" genetic instructions of one period may be the "bad" genetic instructions of another. The ϵ4 allele may be bad now because it increases vulnerability to *C. pneumoniae*, but it may nevertheless be common because *C. pneumoniae* did not make ϵ4 so disadvantageous in the past. *C. pneumoniae* may have used a different infection strategy, one that didn't impose damage on ϵ4 people; or it may not have been present for a long time in populations in which ϵ4 is common. The ϵ4 allele may be associated with increasing vulnerability to *C. pneumoniae* now, but a millennium from now, in response to a lowered frequency of ϵ4 in people, the organism may have changed, and some other instruction might increase vulnerability to the pathogen. Consequently, $\epsilon 4$ can be common now even though it would not be common if it were an inherently bad genetic instruction.

High iron levels are an environmental risk factor for atherosclerosis. High iron levels help explain why women are less vulnerable than men to atherosclerosis before menopause, because menstruation lowers their iron stores. After menopause, iron levels in women increase, and so does their risk of atherosclerosis. Women who have high iron levels are more likely to have atherosclerosis than women who have lower iron levels. The same is true for men. The iron risk factor also helps explain the association between red meat and atherosclerosis because the form of iron obtained in red meat tends to be less suppressible by the body's homeostatic equipment than iron from other sources, such as vitamin supplements. The problem with the iron risk factor arises when mechanisms are considered. According to one suggested mechanism, iron oxidizes fat and that oxidized fat damages the arteries. Evolutionary considerations cast doubt on this sort of explanation, though. With a long evolutionary history of human access to iron and fat, one would expect mechanisms to control this sort of damage, especially given that the vulnerability seems variable in today's population. If this idea were correct, one would have to conclude that protected people are already present. If they had always been present, they should have taken over the population by now.

If one broadens the scope of inquiry to consider pathogens, a more feasible explanation arises. Bacteria need iron for bringing in oxygen and for making some enzymes. The abundant iron in human bodies is largely unavailable to bacteria because it is bound up in molecules rather than being free in solution. Although restricting iron below the normal level reduces the availability of iron for our own purposes, it may starve out bacteria. People with low iron levels may therefore be better able to keep *C. pneumoniae* under control and therefore may suffer less damage than people with high iron levels. The association between iron and atherosclerosis therefore makes more sense in the context of infectious causation than it does if pathogens are not considered.

The other risk factors are similarly better explained and integrated when *C. pneumoniae* is brought into the picture. *C. pneumoniae* is lipophilic (fat-loving)—the accumulation of fat and cholesterol in the atherosclerotic lesions therefore makes sense. *C. pneumoniae* causes oxidation of fats—the association between oxidized fats and atherosclerosis therefore makes sense. *C. pneumoniae* causes inflammation—so the association between inflammation and atherosclerosis makes sense. The positive effect of aspirin on heart disease also makes sense if the inflammation is a self-destructive side effect of infection because aspirin reduces inflammation.

C. pneumoniae develops particularly florid infections in the lungs of smokers. Smoking inhibits and sometimes even destroys the normal defenses of the respiratory tract; it can, for instance, paralyze the hairlike cilia that would normally move pathogen-laden mucus up and out of the lungs. Pathogens can therefore increase to higher densities in the lungs of smokers. The association between smoking and atherosclerosis therefore makes sense, especially the increased risk of atherosclerosis that is associated with secondhand smoke. People exposed to secondhand smoke inhale a tiny fraction of the smoke inhaled by smokers. Yet they have an increased risk of atherosclerosis that is about one third of the increased risk of smokers. This high risk seems way out of proportion to the dosage of smoke but not to the increased dosage of *C. pneumoniae* that could be coughed up from the florid infections in the lungs of smokers. It probably isn't the secondhand smoke that gives you heart disease so much as the greater numbers of pathogens catapulting out of the mouth of a coughing smoker.

Year by year the accumulating evidence is making the problem more interesting, and, for the time being, more complex. Another line of research parallels the work on *C. pneumoniae* but implicates another bacterium. This one, *Porphyromonas gingivalis,* is better known for its role in causing gingivitis and periodontal disease. The evidence parallels the evidence supporting *C. pneumoniae,* but the first incriminating studies were conducted in the early 1990s, about five years after Saikku and Leinonen published their first paper implicating *C. pneumoniae. P. gingivalis,* like *C. pneumoniae,* is found in atherosclerotic lesions, and it can be very destructive when it escapes into the bloodstream from its oral niche. There have now been about ten studies, most of which confirm an association. The researchers who pioneered this work have, like Saikku and Leinonen, found themselves ignored or overlooked, in spite of the great potential importance of their work.

The possibility that two or more pathogens are being implicated as causes of atherosclerosis is no reason to doubt a role for either one. We need only remember hepatitis or pneumonia, each of which can be caused by several entirely different infectious agents. Before the infectious causes of these diseases were identified, they were lumped together into a single category based on similarity of the disease manifestations.

THE PLAGUE OF THE MIND

If we pick a disease that is well understood and look back at those who studied the disease in previous decades, it is easy to see who was on the right path and who was lost in the wilderness. When we look at peptic ulcers, for example, it is easy to see that Connie Guion was on the right path when she told her personnel to treat ulcers with Aureomycin, as was Barry Marshall when he and his colleagues made use of the same treatment in the 1980s. Identifying diseases that will be recognized in the future as being caused by infection is more difficult and of greater consequence. The examples of diseases that have recently been recognized as being caused by infection—liver cancer, cervical cancer, peptic ulcers, and the like—provide an important lesson: their infectious causation has not been recognized by a single critical piece of evidence so much as by a compelling body of evidence. Infectious causation of each of these diseases could still be dismissed by focusing on weak points in a string of logic to discredit the hypothesis. A sufficiently clever critic can transmute a questionable assumption into a shadow of doubt (a craft well honed by well-paid defense lawyers). We are in the midst of renewed recognition of infectious causation of chronic diseases. The recent additions to the list of diseases caused by infection reveal that the building of compelling bodies of evidence is protracted over many years. Considering this course, we should be able to identify those diseases that will soon be recognized as being caused by infection on the basis of how close to compelling the body of evidence has become. The body of evidence for infectious causation of atherosclerosis is approaching the zone of compelling evidence. For others, this compelling zone is undoubtedly much more distant.

Atherosclerosis would be enough to make *C. pneumoniae* rank among the most deadly human pathogens of all time. But *C. pneumoniae* appears to be playing a role in other diseases. One of the most important of these potential roles is being investigated by Alan Hudson and his colleagues at Wayne State University. Hudson is no stuffy academic. The slightly disheveled gray-white hair brushing his shoulders makes it clear that he is not a status-conscious medical type glancing in a mirror to check out his respect quotient. He is up front and gets down to business; and his business is conducted with an unmistakable delight in what the tools of modern science can reveal. His ideas shoot out in volleys as he navigates from project to project, introducing each with a preface such as, "Oh, you'll love this!" Then he's off with his explanation. In a conversation over a beer in a Detroit bar and grill, he first talked about his work implicating *C. psittaci* in temporo mandibular jaw syndrome—this association, along with the tendency for women to own pet birds, finally offers an explanation for the tendency of TMJ syndrome to be particularly common in women. Before I digested that, he had gone on to his work implicating *C. pneumoniae* in the inflammation of arteries (arteritis). Then he was on to a side project on the transmission of an asymptomatic *C. trachomatis* infection: a woman who had claimed no other sexual contacts besides her husband had somehow acquired a florid *Chlamydia* infection, yet her husband had shown no symptoms since receiving treatment for a *Chlamydia* infection many years earlier, before they met. A sense of trust and perhaps a marriage were hanging in the balance; each had the uneasy sense that the other might have been having an affair. Hudson explained the resolution with his characteristic enthusiasm: "We PCRed the guy for *C. trachomatis*. His test came back ferociously positive!" Even the hardened antiscience resolve of a deconstructionist might melt away after an hour of conversation with Alan Hudson.

Hudson was delighted that his molecular tools might provide insight into the dangers of asymptomatic infection and save a marriage to boot. But one of his projects has far broader implications. The work started with discussions he had with a neuropathologist, Brian Balin, in Philadelphia while the two scientists were fulfilling the arduous and unpleasant duty of service on a disciplinary committee prior to Hudson's move to Wayne State. The committee, evaluating allegations of misconduct of a Ph.D. student, met repeatedly over a period of several months. To break the tedium of the seemingly endless meetings, Balin and Hudson would strike up conversations about scientific work at the interface of their disciplines—Balin's neuroscience and Hudson's molecular biology and microbiology. Stimulated in part by Saikku and Leinonen's research on atherosclerosis, Balin asked Hudson whether *Chlamydia* might also be involved in Alzheimer's disease. Hudson was skeptical, but, knowing that the sexually transmitted species of *Chlamydia*, *C. trachomatis*, sometimes invades the body and causes arthritis, he agreed to test some samples. Balin sent him about thirty human brains. About half of them were from people who had died of

Alzheimer's; the other half were from people who did not have Alzheimer's at the time of death. All of the brains were coded so that workers in the Hudson lab would not know which was which. None of the brains tested positive for *C. trachomatis.*

Hudson then asked his post-doctoral fellow, Hervé C. Gérard, to test the brains using a new, very sensitive molecular test that the lab had just developed for detecting *C. pneumoniae* in arthritis patients. Half of the brains came back "ferociously positive" for *C. pneumoniae.* Hudson has nothing but praise for the technical skills of Gérard, but, just to be sure—and to cope with his sense of astonishment—he told Gérard to test the samples again. The results were the same. When the code was broken, Hudson found that almost all of the infected brains were from Alzheimer's patients. As of May 2000, Hudson's lab has tested twenty-three brains from severe cases of Alzheimer's disease, and twenty-five "control" brains from people who had no signs of Alzheimer's before they died. All but one of the brains from Alzheimer's patients were positive for *C. pneumoniae;* only one of the control brains was.

Like the hypothesis of infectious causation of atherosclerosis, the hypothesis of infectious causation of Alzheimer's disease unifies seemingly unrelated evidence. Most important, it unifies the information on atherosclerosis with a separate disease that may be better understood as a different aspect of an underlying infectious process. The link is apparent at the level of host genetics as well: the $\epsilon 4$ allele is a risk factor for Alzheimer's disease as well as for atherosclerosis. It was Hudson's lab that did the clever experiment on arthritis patients which indicated that the $\epsilon 4$ allele conferred susceptibility to *C. pneumoniae.* The experiment was planned with Alzheimer's in mind.

The growing body of evidence suggests that atherosclerosis and Alzheimer's disease—two of the most common and damaging chronic diseases—may be added within the next decade or so to the growing list of diseases caused by infection. These imminent additions, together with the discoveries of the last two decades of the twentieth century, leave one wondering just how lopsided the comparison will be between the old infectious killers that have been newly recognized and the infectious killers that have newly emerged as global scourges. At present there is just one disease that falls into the second category, and it is the one that is most responsible for directing medical attention toward infectious causation of chronic disease: AIDS.

CHAPTER NINE

Diseases of Blood and Steel

WHAT ABOUT THE CHRONIC DISEASE THAT SPARKED THE CURrent concern over emerging diseases? The conventional wisdom is that AIDS arose from a secluded area in Africa where a virus was transmitted to humans from another primate, spread locally, and then spread globally within a decade or so. If this interpretation is correct, the AIDS pandemic, together with the lack of similar novel pandemics throughout the past two centuries, gives some perspective on the threat of new pandemics over the next few centuries. The threat would be real over the long run but probably not imminent over the short run. Yet according to a different hypothesis for the origin of AIDS, most thoroughly presented by Edward Hooper in his book *The River,* the pandemic may have been more controllable than is generally believed. The control in this case would not have been generated from a surveillance system or an ability to combat the pathogen after it was globally embedded. Rather, the control could have been generated by better safeguards of medical procedures.

Resolving the question of the origin of AIDS bears on the broader theme of where the future threats to human health will come from. AIDS is a chronic malady. The arrival of the AIDS pandemic therefore accords with the idea that the stealth infections are the modern threats for the most prosperous countries; but the various explanations for the origin of AIDS implicate different social, geographic, and biological sources of the AIDS pandemic. AIDS is the only twentieth-century example of a new kind of pandemic plague; its emergence therefore provides the only hard evidence of how a pathogen has emerged—and therefore can emerge—in modern times from a local source to cause a fundamentally new kind of pandemic plague. If we ignore this bit of hard evidence, we risk our future.

Hooper argues that the contamination of polio vaccines used during the late 1950s in sub-Saharan Africa is the source of the AIDS pandemic. This idea has surfaced in several places since the early 1990s. It has been dismissed by many experts, often with ridicule. But ridicule from experts is not necessarily a good reason for rejecting an idea. To move toward the truth, we need to assess the evidence.

Viruses need to grow in host cells. So to generate polioviruses in sufficient abundance for use in polio vaccines, virologists grew the viruses in cells taken from other primates, such as green monkeys. The cells and viruses were grown together in flasks containing media suitable for cell growth. After the viruses had replicated to high densities, they were separated from the monkey cells, media, and cellular debris left in the media by the viral feast. The viruses were then used in the next phase of vaccine preparation, and everything else was discarded. Ideally only the poliovirus would be isolated during this step, but the laboratory procedures that separate polioviruses at this stage often fail to exclude other viruses that might have been living in the cells that were taken from the monkeys and that reproduced in the cell culture right along with the poliovirus. Without adequate safeguards, these viruses could inadvertently be harvested with the polioviruses and included in the vaccine as contaminants.

Just this kind of contamination of polio vaccines occurred during the 1950s and early 1960s. A particularly worrisome contaminant, a virus called SV40 (for simian virus number 40), was inoculated into children along with the killed polio vaccine virus. The procedure used for killing the poliovirus during the preparation of the Salk vaccine did not kill all of the SV40s. The concerns over contamination of oral polio vaccines have generally been even greater because oral polio vaccines introduce live viruses and so were not subjected to treatments that inactivate poliovirus and might similarly inactivate other stowaway viruses. Indeed, some viruses have been found in live polio vaccines, including at least one retrovirus, the family of viruses to which HIV belongs.

The transmission of SV40 in polio vaccines lends some credibility to the possibility that a precursor of HIV could have similarly contaminated cells used for the production of polioviruses. This "contaminated vaccine" hypothesis offers a mechanism for what looks like a simultaneous transfer of different immunodeficiency viruses to humans from simians (nonhuman primates such as monkeys and chimpanzees). The names used for these viruses can be confusing. When one of them is isolated from humans it is called HIV (human immunodeficiency virus). When one is isolated from a simian (a monkey or chimpanzee) it is called SIV (Simian Immunodeficiency Virus). HIV experts believe that different HIVs have arisen from the transfer of different SIVs from simians to humans. The pandemic HIVs belong to one half of the HIV evolutionary tree and are referred to as HIV-1s. The pandemic HIVs make up the main cluster of viruses in the HIV-1 tree and are referred to as group M, M being short for "main." Most of the other HIV-1s are found in West Central Africa, particularly Cameroon and Gabon; they are collected into a group called O, for "outgroup." The other half of the HIVs arose in West Africa and are referred to as HIV-2s. The HIV-1s are most similar to a group of SIVs found in chimpanzees; the HIV-2s are most similar to a group of SIVs found in sooty mangabeys.

Evolutionary relationships among the different HIVs and SIVs can be deduced by comparing the sequences of building blocks that make up the genes of the different viruses. All else being equal, the more similar the genetic sequences of the viruses, the more recent is the last common ancestor shared by the two viruses. These evolutionary trees allow researchers to visualize branching points between different genealogical lineages of viruses. When these family lineages are combined with information about the species from which the viruses were isolated, researchers can speculate about when viruses may have moved from one species to another. The more complete the evolutionary tree, the more reliable the interpretation. Some researchers using this method suggest that several viruses recently entered humans from simians over a short period of time, perhaps a decade or even less.

If this interpretation is true, it raises a perplexing question. Why would immunodeficiency viruses stay put in simians for thousands of years and then suddenly, within a decade, be transferred several times into humans? The contaminated-vaccine hypothesis offers an explanation: cells used for oral polio vaccines were contaminated with several different SIVs that were transmitted to humans during the first decade of the oral polio vaccination program, thus generating several different lineages of HIVs.

Some incarnations of the contaminated-vaccine hypothesis can be dismissed on the basis of available information. Because monkey kidney cells were the standard for growing polioviruses, and SIVs were present in green monkeys, early attention, prompted by New Hampshire attorney Walter Kyle, focused on this possible source of HIV. Critics rightly pointed out that the SIVs found in green monkeys were not the viruses implicated in the origin of HIV. The genealogical trees instead implicated SIVs found in sooty mangabeys for the origin of HIV-2, and SIVs found in chimpanzees for HIV-1. Hooper and other defenders of the contaminated-vaccine hypothesis pointed out that many different primates may have been used as sources for cells, perhaps even chimpanzees and sooty mangabeys. Vaccine researchers in sub-Saharan Africa may have used locally available primates—chimpanzees in one area, green monkeys in another, and mangabeys in still another. This potential for local use confounds explanations of natural origin, which often assume that the presence of similar viruses in humans and nonhuman primates of a given area implies transmission from the primates to the humans by natural means. The contaminated-vaccine hypothesis suggests that the same pattern may arise from artificial transmission because cells from indigenous animals were used to prepare the vaccines of a particular region.

Critics also pointed out that kidney cells were not the right kind of cells for primate immunodeficiency viruses, but defenders responded that the right kind of cells—white blood cells—are often present in cultures of kidney cells that are used in vaccine preparation. White blood cells containing the SIVs could therefore contaminate vaccine stocks made from kidney cells.

The contaminated-vaccine hypothesis was given added weight when the great evolutionary biologist William D. Hamilton emphasized the need to evaluate it more rigorously in the foreword to Hooper's book. The mild-mannered Hamilton had a long track record of putting forward groundbreaking ideas without fanfare, being overlooked, and then, over the ensuing decade or so, being borne out by the evidence. Evolutionary biologists have learned to listen carefully to his soft-spoken assertions. Hamilton did not claim that contaminated vaccine lots introduced a simian virus into humans to create HIV. Rather, in keeping with principled guidelines of scientific inquiry, he suggested that this idea was being too eagerly dismissed without adequate evidence.

The contaminated-vaccine hypothesis would be particularly compelling if the diversification of the HIV-1 lineages had occurred just after the widespread use of the vaccine, and if there were no evidence of transmission to humans before or after this time. In this case only one contamination event would be necessary and the perplexing question "Why now?" would have a distinct answer: inadequate quality control and bad luck.

Reconstructions of HIV's evolutionary genealogy, however, have generated a different result. By estimating the rate of change of HIV-1 genetic sequences, researchers now conclude that the first diversification of HIV-1 lineages occurred two or more decades before the use of the polio vaccine. If that diversification occurred in humans, the transmission to humans would have occurred more than a decade or two before the oral polio vaccine program. The diversification of the ancestors of HIV into the HIV-1 and HIV-2 halves of the evolutionary tree would have occurred long before that, perhaps hundreds or even thousands of years ago.

This prevaccination divergence of the HIV-1 lineages weakens the contaminated-vaccine hypothesis for the origin of HIV-1 but does not deal it a deathblow. A still viable possibility is that the early diversification occurred in chimpanzees rather than in humans, and the oral polio vaccines were contaminated by several different chimpanzee viruses. According to this idea, each of the several main (group M) HIV-1 lineages would descend from a different transmission event to humans.

If this idea is true, then some of the descendants of those chimp viruses never left chimpanzees, and we might still be able to find them there, in chimpanzees. All known chimp SIVs, however, branch off from the genealogical tree before the diversification of the group M HIV-1s—that is, all except for two fascinating viruses that were recently isolated from two chimp brothers caught in the wild. These two viruses are closely related evolutionarily. Each is a recombinant virus made up partly of genetic material that resembles HIV-1s of group M and partly of genetic material that resembles the chimp SIVs as a whole as much as it resembles the HIV-1s as a whole. There are four feasible explanations for these viruses: (i) they descended from a chimp virus that entered humans, recombined with a human virus, and then went back into chimps; (ii) they descended from human viruses that recombined in humans and then went into chimps; (iii) they descended from a human virus and a chimp virus that recombined in chimps after the human virus was transmitted to chimps; or (iv) they are descended from chimp viruses, of the kind suggested by the polio vaccine hypothesis, that recombined in chimps.

Whichever of these hypotheses is correct, the conventional understanding of the HIV-1 pandemic will be challenged. The first three hypotheses are iconoclastic because they propose that some chimp viruses have evolved from human viruses. If so, then perhaps other SIVs have resulted from human-to-chimp transmission, and perhaps HIV has a longer history in humans than is generally believed. The fourth possibility is striking because it implies a high rate of transmission to humans at some time many decades ago. Such a flurry of transmission requires some temporarily opened avenue, such as contaminated polio vaccines.

As Hamilton pointed out, these ideas can be tested by looking to see what strains are now present in what species and in any remaining vaccine stocks. A better sampling of SIVs from chimps, and HIVs from sub-Saharan West and Central Africa, should resolve much of the uncertainty by showing whether these intriguing chimp SIVs cluster within the HIV genealogical tree, implicating human-to-chimp transfer, or whether the different lineages of human viruses cluster within the genealogical tree of chimp viruses, implicating a sudden transfer to humans.

Sampling of human viruses in West Central Africa is lending credence to the scenarios that involve human-to-chimp transmission. One group of viruses isolated from northern Cameroon, Nigeria, and Djibouti is a mosaic of HIV-1 subtypes, resembling the kind of virus that could have served as an ancestor to one part of the mosaic chimp viruses. These and other group M subtypes are known to recombine in infected people with the more archaic group O subtypes of HIV-1, which are the kinds of virus that would be needed as a precursor for the other part of the mosaic chimp viruses.

Hooper also proposes that the HIV-2 infections in West Africa could have resulted from transmission to humans via a polio vaccine, though in this case the vaccines would have been contaminated with viruses from sooty mangabeys rather than chimpanzees. The mildness of HIV-2 relative to HIV-1 provides an opportunity for testing that is not available for most strains of HIV-1. Because of this mildness, people who might have been infected before the administration of the polio vaccine could still be alive and infected. An analysis of the prevalence of infection in different age groups could therefore reveal a statistical footprint. If people were first infected with HIV in the late 1950s from a polio vaccine, then sexually active people who were older at that time would not have had a longer exposure to HIV infection than sexually active people who were younger at that time. In contrast, if in the late 1950s HIV had already been endemic in the population, the older sexually active people should have had a higher prevalence of infection than the younger people. This statistical footprint left in the age-related pattern of infection should be apparent in assessments of infection conducted in subsequent decades. The earlier the collection of data, the more distinct the footprint.

The earliest batches of data were collected in the late 1980s. Fortunately, this date of collection is just early enough to provide a test, particularly if the strains are very mild. Hooper placed considerable emphasis on this logic as he considered the prevalence of HIV-2 infection in different age groups in the West African country of Guinea-Bissau. The peak prevalence in Guinea-Bissau, which occurred among people 40–50 years old, and the rapid decline in older groups is consistent with a transfer of SIV to humans through contaminated polio vaccines. It is also, however, consistent with an increased rate of spread during this time from other causes, such as social turmoil, which could have increased the rate of infection, or from entrance of HIV-2 into Guinea-Bissau from a neighboring country during that period.

Evidence from other countries in which HIV-2 is endemic may provide insight. If no other countries showed a prevalence that continued rising among people who were old enough to have been sexually active before the time of the vaccine (people who were older than 50 in the late 1980s), then the HIV-2 experience as a whole would be consistent with the contaminated-vaccine hypothesis. If, however, one found a rise that continued through age classes of people who had been sexually active before the time of the vaccine, then one would have grounds for rejecting the contaminated-vaccine hypothesis. The test would be most powerful in countries that were home to particularly mild HIV-2 infections, because mild infections that originated before the administration of the polio vaccine would have a better chance than severe infections of being around three decades later to leave a footprint of their prevaccine existence. Studies of the associations between HIV-2 infection and AIDS indicate that infections in Senegal best meet this criterion. Most people who are infected with HIV-2 in Senegal will probably not die of AIDS. If HIV-2 first infected people in Senegal via the polio vaccination, people who were around 30 or 40 at that time should not have a higher positivity rate than people who were 20 at the time. Considering this population thirty years later, one would predict that people who were 60 or 70 in the late 1980s would not have a higher infection rate than people who were 50.

In contrast with this prediction from the contaminated-vaccine hypothesis, HIV-2 infections during the late 1980s increased steadily with increasing age. HIV-2 appears to have been endemic in the Senegalese population well before the first use of the polio vaccine.

If HIV-2 first entered humans in 1958, we would expect that people whose sexually active history dates to earlier times should not be increasingly infected with HIV-2. Specifically, of the women sampled between 1985 and 1990, those who had been sexually active for more than thirty years (that is, before the use of the polio vaccine) should not be infected with HIV-2 more frequently than women who had been sexually active for just thirty years. Again, the data from Senegal do not support the contaminated-vaccine hypothesis—the longer the history of sexual activity, the higher the prevalence of infection, even among people who had been active for more than thirty years. The associations between age, sexual activity, and HIV-2 infection in Senegal therefore argue against contaminated polio vaccines as the origin of the HIV-2 epidemic.

One could still argue that HIV-1 lineages were derived from contaminated polio vaccines. But once the discordance between the HIV-2 experience and the vaccine hypothesis has been granted, the need for the polio vaccine hypothesis vanishes—HIV-2 provides the evidence that some other route of entry into humans has occurred. Still, to resolve the issue, strains from humans and chimps must be isolated and evaluated to determine whether the anciently branching lineages implicate humans or chimps as the source of HIV-1s and to assess the timing of the transfers.

The more anciently derived diversity occurs in West Central Africa, in countries such as Cameroon, Gabon, and Equatorial Guinea. That is where a wide range of strains would have to have been introduced by vaccines to have generated the divergent lineages of HIV-1. Consideration of that region is surprisingly lacking in Hooper's analysis, even though the region is critical for evaluating whether HIV could remain smoldering over long periods. Unfortunately it is probably too late to assess age prevalence data among people in that area, in the same way that we have just assessed it for HIV-2, because people who are in their 70s, 80s, and 90s are probably not having enough sex to put them at risk for new infections. Although far fewer isolates of these viruses have been studied, the diversity of their genetic sequences is comparable to the diversity found among the pandemic group M viruses.

Did these viruses generate this diversity over similar short periods of time? Or have the divergent viruses, which have clearly not infected people at a comparable rate, accumulated this variation much more slowly? The slower generation of variation seems the more reasonable explanation, because the vastly greater amounts of replication among the group M viruses during their pandemic spread should have generated vastly greater diversity in genetic sequences. Evidence to evaluate this idea can be obtained from early infections among people who were geographically isolated from the area. A Norwegian traveler was infected in the 1960s with a group O HIV-1 that looked very much like the recent isolates of the viruses. This bit of evidence indicates that the O group viruses have been evolving much more slowly than the pandemic M group viruses. That slow evolution in turn indicates that the diversification of the O group viruses began long before the polio vaccine was introduced. To rescue the vaccine hypothesis as an explanation for the O group of HIV-1s, one must again presume that vaccine lots were contaminated with different lineages of chimp viruses that then served as ancestors to the different lineages of O group viruses.

Although the available evidence cannot definitively exclude the contaminated-vaccine hypothesis, the evidence has been steadily restricting its zone of feasibility. That restriction makes the critical test of the hypothesis easier to envision. If the contaminated-vaccine hypothesis is correct, certain SIV strains in chimpanzees would match up with certain viruses of the major lineages of HIV-1 within the O group and within the M group, those matches being much more similar than the SIVs are to each other, and much more simliar than the different major lineages of HIV-1 are to each other. The few strains isolated from chimps already show a great degree of genetic diversity, great enough to indicate that the lineages predicted by the contaminated-vaccine hypothesis would still be represented in the chimp populations. Considering the large numbers of group O–like SIVs and group M–like SIVs that would have been in chimps according to the contaminated-vaccine hypothesis, descendants of these proposed ancestors should be left in chimps waiting to be detected.

One obstacle to this detection is the difficulty of sampling chimp populations without harming them. There may be another way, however. Chimpanzee droppings are laden with the cells of the chimpanzee. If some of those cells are infected with SIVs, the SIVs will be present in the chimpanzee dung. If the SIVs can be extracted from the rest of the mess and their genes amplified, their genetic sequence could be obtained. Researchers might then be able to resolve this thorny issue by going to Central and West Central Africa to collect chimpanzee dung.

When I began writing this chapter in January 2000, W. D. Hamilton had gone to the Congo to search for such viruses. By the time I had finished the first draft, he had been whisked back to a London hospital with what was apparently a raging case of malaria. He seemed to be recovering but then fell deathly ill and unconscious. On March 7 the discipline of evolutionary biology and the enterprise of science lost one of its deepest minds. A balanced resolution of this problem, whichever way the evidence points, would be an appropriate tribute to a man whose life was dedicated to the truest principles of free and thoughtful scientific inquiry. The balance of evidence now seems to be against the contaminated-vaccine hypothesis, but the more detailed analysis of the chimpanzee viruses that Hamilton sought is still needed to resolve the matter.

If HIV did not enter humans from contaminated vaccines, then how can one explain the apparent simultaneity of the emergence of the different HIV groups in humans? One possibility is that the simultaneity is more apparent than real. Different HIVs have evolved sequence changes at dramatically different rates. The influences of these different rates and the free play afforded by the gaps in the evolutionary trees lead to the conclusion that the entrance of different HIVs into humans could have been spread out over many decades or even centuries. Contaminated vaccines were suggested as a cause for the HIV pandemic partly because it was difficult to envision how HIV had apparently entered humans almost simultaneously over a decade or so in the mid-twentieth century but had failed to do so during previous centuries. The steadily rising prevalence of HIV-2 infection in older age groups and among people who have had more decades of sexual experience severely weakens the vaccine hypothesis for HIV-2. The genealogical evidence for a prevaccine divergence of HIV-1 strains, along with the slow generation of diversity among some HIV-1 lineages, similarly weakens the vaccine hypothesis as a general explanation for the generation of

HIV-1. In so doing, this evidence removes the reason the vaccine hypothesis seemed compelling. It was attractive because it could explain the seemingly simultaneous emergence of several HIV lineages. Now it seems that we do not need a novel mechanism to account for what appeared to be a nearly simultaneous entry of all the HIVs into humans.

So what is the alternative? Although one may not need a novel mechanism for simultaneous transmission of many different HIV lineages to humans, one must still explain why the AIDS pandemic occurred now rather than in centuries past. One possibility is that the precursors to the pandemic HIVs of group M were smoldering along in secluded groups of people. The natural follow-up question is, "So why didn't they spread?"

The best answer has two parts. The first part is theoretical. When a virus first gets into humans, it may take a while for it to become adept at infecting humans and being transmitted from humans. Before it develops such abilities it may just barely get by. Depending on the biochemical innovations that may be necessary, this process could take a long time, perhaps centuries if some new structure, like a new binding protein, is necessary. Even after the innovation develops, the biochemical machinery needs to become tuned to the individual host, and the host population would need to have the right characteristics for pandemic spread of the new variants.

The second part of the answer draws attention to the pattern of evidence that one actually sees in retroviruses in humans but does not get bogged down in the theoretical reasons for those patterns. The general pattern among human retroviruses is borderline persistence without explosive pandemic spread. We have, as everyone knows, one case of explosive pandemic spread: the group M HIV-1s. The group O HIV-1s are composed of smoldering viruses. This group includes a few stray global wanderers: in Los Angeles, Spain, France, Belgium, and a few other places outside West Central Africa. Considering the tremendous potential for world travel, one would expect some such wanderings.

But these wanderers have not sparked new foci of epidemics as did the wanderers that belonged to the group M HIV-1s. Group O viruses have remained largely endemic in West Central Africa, apparently for at least a century and possibly for many centuries. One other group of HIV-1s, called group N, appears to be a relic of a similar HIV-1, one with even less diversification and persistence than the group O viruses.

The geographical history of HIV-2 is similar. Two of the HIV-2 groups have maintained themselves largely in West Africa, with a few global wanderings, mainly to Europe. But HIV-2 has not caused major epidemics outside the African continent. Like group N HIV-1s, some other lineages of HIV-2 are represented by only one or a few isolates. These genealogical stragglers look more similar to sooty mangabey viruses than to other HIV-2s but are moderately distinct from either. Their uniqueness could have resulted from their just getting by in humans for many transmission cycles—all or nearly all of the viruses to which they are closely related having been lost over time. Or their uniqueness might be more apparent than real, a result of inadequate sampling of sooty mangabeys for viruses. Whichever is the correct explanation, none of these straggler HIV-2 lineages appear to be in the process of generating anything like an explosive outbreak. Rather they appear to be smoldering slowly or barely persisting in human populations.

When we look to HIV's cousins in the human matrix, the HTLVs, we see the same combination of viruses that are slowly creeping along or clinging to persistence. Unlike the HIVs, however, the HTLVs show the consequences of this predicament over a more expansive period of time. HTLV has probably been in humans for tens of thousands of years, moving where humans moved. Its heartland, like that of HIV, was Africa. Over this longer period of time it has seeded different parts of the earth and set up outposts of infection. One is among African Americans in the southeastern United States, apparently a legacy of the HTLV-I that was inadvertently brought to the New World in Africans during the slave trade. Another is in Japan, where the smoldering pockets of infection led researchers to realize that the leukemia it causes must have an infectious origin. Some communities had a distinctive leukemia that was virtually absent just a hundred miles away. HTLV-II appears to have had a similarly slow spread out of Africa and into the New World many thousands of years ago, this time overland through the Bering Strait when the last ice age made the strait into a bridge. One of its legacies is its presence in isolated populations of American Indians. Only recently have these HTLV-IIs broken out from this confinement and into intravenous drug users. In the Old World HTLV-II is left as a remnant in indigenous people of West Central Africa commonly known as Pygmies.

When people discussing the AIDS pandemic ask "Why now?" they imply that some special process must be invoked to explain its timing. But the HIV/SIV genealogical tree and the distribution of the different kinds of HIV in circulation today are consistent with introductions from simians to humans repeatedly over a much longer period, some taking hold but not spreading, and others not taking hold. By this reasoning we are left today with one introduction that took hold and spread globally, generating the HIV-1 pandemic, and several other HIVs that either took hold locally or spread only slightly in different areas of West Africa. There may also have been many that have hardly taken hold at all, and may be lost within the next few years. We see at any time only those that have not been lost.

In the genome of humans and simians are embedded endogenous retroviruses, relics of viruses that may have entered our ancestors millions of years ago and are now passed along from generation to generation in sperm and egg cells. These endogenous retroviruses reveal that retroviral infection has occurred many times in the past, but they do not tell us how frequently. New retroviral infections need a very specific attribute to become endogenous retroviruses—they must become integrated into the DNA of our germ line, our sperm and egg cells. We do not know what proportion of human retroviruses would eventually find their way into the germ line. If there are ten thousand retrovirus invasions into humans from some other host species for every one that becomes endogenous, then our evolutionary history might have involved a continual rain of retroviruses that infected our ancestors and then disappeared without a trace. Those HIVs and HTLVs that concern us today may be just the most recent raindrops on the human species, perhaps persisting now slightly more than before because there are more people providing more opportunities for transmission than ever before.

The problem is much like considering possibilities of intelligent life in the universe by looking only at earth. By looking only at our planet, we cannot deduce whether the universe is teeming with intelligent life, devoid of intelligent life, or anything in between. We have only one data point, earth, and cannot generalize about the rest of the universe; we do not know how likely or unlikely it is that the conditions leading to intelligent life have been repeated throughout the universe. So too, we have only one data point for a rapidly spreading retroviral pandemic, and we do not know how likely or unlikely it is that the conditions leading to this situation will recur through time. It could be one out of ten retroviral introductions, or one out of ten thousand.

The current state of knowledge does not require the identification of some special process to explain the current array of HIV pandemic, epidemic, and endemic infections. Retroviruses in humans are characterized by slowly smoldering spread, partly because it may take a long time for a sexually transmitted virus to generate in a new host species innovations that permit a high degree of transmissibility while avoiding immunological destruction. When such a change occurs, large-scale transmission is possible if the options for capitalizing on that transmissibility are present—that is, if there is a high potential for sexual transmission within the population, and between populations in ever widening spheres of infection. If such opportunities are not available, the new variant, with all its pandemic potential, cannot spread pandemically; it may even vanish if the transmissibility comes at too great a cost—for example, if increased densities of viruses in semen result from increased exploitation of the host, which in turn shortens the host's life span by causing AIDS. If, however, the innovation that allows increased host exploitation co-occurs with golden opportunities for transmission, then the new variant can spread rapidly out of the local pocket and keep on spreading globally. According to this scenario, only the group M infections have had this conjunction of innovation and opportunity, and thus only the group M viruses have broken out of the typical retroviral mold and assumed a pandemic form. Whenever such a conjunction of events happens, we might look and wonder why. But our answer is both probabilistic and deterministic—the conjunction depends on both the social setting and chance. In the case of the AIDS pandemic the chance event—the generation of a more highly transmissible and exploitative variant—occurred in people who were linked in a sexual network with populations that had a high potential for ever expanding spheres of sexual transmission, and so the pandemic occurred. The other retroviruses have not yet had that dangerous conjunction.

They could be nowhere near this conjunction, or they could be dangerously close. We do not yet know enough about the process, but we can keep our eyes open for the retroviruses that look closer to it than the others. At the beginning of the 1990s I cautioned that HTLV-II was at the brink of such a confluence of events. I singled out HTLV-II because it was a relatively benign virus that had spread from its source in often remote groups of American Indians into the intravenous-drug-using population. Its entrance into intravenous drug users provided opportunities that could favor transmission of particularly dangerous variants. The main question was whether HTLV-II would undergo the dangerous genetic changes that would be favored by this opportunity. HTLV-II has not exploded yet, but an ominous sign has recently been reported. It looks as though its replication rate has increased: its timer is ticking faster. Belgian investigators assessed the rate at which HTLV-II has been evolving in intravenous drug users by comparing its rate of genetic change in that population with the rate that has been occurring in indigenous people of West Central Africa and American Indians. In these groups the virus is spread from mothers to children through breast milk, probably also augmented by some sexual transmission. The rate of viral evolution in intravenous drug users was 150 to 350 times faster than in the other groups.

This is a trend worth worrying about. It is reminiscent of the unnoticed warnings of another epidemic. Retrospective studies of blood taken in the early 1970s showed that HIV was infecting about one out of every hundred injecting drug users in U.S. cities, years before it was common in homosexual men. By the time the outbreak was first being noticed in gay men, about one out of three injecting drug users in New York City was infected.

How many times during the 1970s was illness in a sickly injecting drug user dismissed as a direct effect of addiction instead of an effect of pathogens that were hitching rides on syringes and needles? How many times have HTLV-II infections, which are normally considered to be benign, caused damage that is similarly dismissed as some effect of addiction? During the 1990s health care workers acquired the AIDS label to explain some of this ill health. It took several years for the HIV outbreak occurring among injecting drug users to develop its counterpart in other sectors of society. Will HTLV-II follow suit? If so, how long will it take? Has it already become more damaging as a result of its evolutionary changes in injecting drug users? Efforts to answer these questions would be a wise investment.

THE SV40 WATCH

Although the available evidence cannot exclude with certainty contaminated polio vaccines as the source of the AIDS pandemic, the balance of evidence argues against it—contaminated vaccines do not appear to have been responsible for what is shaping up to be one of the worst pandemics in human history. Does this conclusion mean that we do not need to worry about vaccines as a source of emerging disease? No. Vaccine contamination has already shown its ability to cause outbreaks of nasty viral diseases. As the United States was playing catch-up during its first year in World War II, a yellow fever vaccination program inadvertently infected over a quarter million soldiers with the hepatitis B virus. Fifty thousand became sick, but they got a lucky break. Apparently, when the hepatitis B virus infected people through this vaccine, it generated a fantastically low number of carriers and thus almost no chronic hepatitis and no liver cancer. Otherwise the vaccination might have caused around ten thousand deaths among the soldiers, and perhaps many more as the virus continued to spread through their contacts.

Now, more than a half century later, vaccine researchers are still nervously watching a contaminant of polio vaccines: simian virus 40 (SV40), the virus that lent credibility to the polio vaccine hypothesis for the origin of AIDS. SV40 naturally infects the monkey cells in which the polio vaccine viruses were grown. Safety experts knew it was there but were not particularly concerned about it because it was not known to be harmful to humans. Their level of concern changed in the early 1960s when experimental studies showed that the virus could cause lung and brain cancer in hamsters. The virus was expunged from the vaccines by the mid-1960s, but that was too late for those of us who were vaccinated between 1955 and 1963. In the United States "those of us" make up about one third of the population. No one knows yet just how bad SV40 is. We are now the experimental hamsters being studied to obtain the answer. We do know that SV40 is found in most cases of a rare, dangerous lung cancer called mesothelioma, which has been increasing in the United States, Britain, and Europe since the use of SV40-contaminated vaccines. This is the same kind of lung cancer that SV40 causes in hamsters. The virus sabotages the same defenses against cancer that are sabotaged by papillomaviruses: p53 and a protein known as retinoblastoma suppressor. The emerging consensus is that SV40 is not just an innocent bystander but is causing the human mesotheliomas in concert with asbestos. SV40 is also found in human brain tumors. Whether it causes brain tumors is still being debated, but its activity in these cells is not comforting: as in the mesotheliomas, SV40 was found in human brain tumors bound to the cells' tumor-suppressing proteins.

The role of SV40 in human cancer will become more apparent as the experiment on the human hamsters continues. The most recent studies indicate that those who received the vaccine are about twice as likely to develop mesothelioma; the risk of the particular brain tumor most associated with SV40 appears to be increased by about one third in vaccine recipients.

If SV40 does not get transmitted from person to person, then the cancer caused by SV40 contamination of polio vaccines can be viewed as another it-could-have-been-worse lesson imparted by the vaccine practices of the twentieth century. But the potential for transmission of SV40 between people is unclear. It has been recovered from children and HIV-infected patients who were born after SV40 had been purged from polio vaccines. The source of these SV40 infections is unknown, but if the virus is transmissible from person to person, our lesson may be just beginning. Concern is heightened because SV40 is a polyomavirus, a kind of virus that includes two other members, called JC and BK, which are full-fledged human pathogens. Most people are infected, for example, with JC polyomavirus, which resides in the brain and other organs. The scope of its effects are not yet clear, but it has recently been linked to colon cancer.

Though vaccine contamination is frightening, in the long run it will probably turn out to be preventable, as soon as the people responsible for vaccine safety know what infectious contaminants to look for. This goal could be reached in time to benefit our grandchildren or great-grandchildren. In the meantime, we can expect that molecular techniques for identifying unwanted hitchhikers will provide incremental advances toward this goal.

ORIGINS

Many inside and outside the health sciences have wondered, "Why all the fuss about determining the origins of the AIDS pandemic and other outbreaks?" Resolution of where things come from guides decisions about our future, ultimately enabling us to put safeguards in place. Although there is a great deal of concern about the danger of emerging new diseases, the most important incident justifying this attention has been the AIDS pandemic. If it turns out that the AIDS pandemic resulted from polio vaccination, then we have no example from the twentieth century of a new kind of pandemic blindsiding society by arising from a secluded source. Nor do we have an example from the nineteenth century. The closest we can come are recurrences of known diseases such as influenza and cholera. To find a pandemic of a previously unrecognized disease caused by a previously secluded pathogen, we would have to go back to earlier centuries, to the first spread of exotic diseases such as yellow fever, cholera, and syphilis.

If, on the other hand, we can conclude that the AIDS pandemic was caused by a natural process rather than a medical bungle, then we will know that we should be ready for similar occurrences over the next few centuries. This recognition, however, should not distract us from the horrible infectious plagues that were here among us during the decades before the AIDS pandemic and that are still here. We have neglected them because we failed to recognize their infectious character. And we still do.

CHAPTER TEN

Modern Miasmas

"MIASMA" IS FROM THE GREEK *MIAINEIN*, MEANING "TO POLLUTE." Medical experts gave the term its notoriety, particularly during the nineteenth century, when they dismissed the idea of invisible pathogens in favor of miasmas: vaporous exhalations from patients or the environment that were assumed to cause disease.

In *The Conquest of Epidemic Disease*, Charles-Edward Amory Winslow remarked, "It is fascinating to note how, for two millennia, laymen were generally contagionists and physicians were miasmatists." Winslow's observation was perceptive, but his rationalization was not: "The physician, knowing more, was quite correct in denying that any *then-available* theory of contagion could explain the facts." Actually, the theory of contagion available throughout the nineteenth, eighteenth, seventeenth, and most of the sixteenth century, could explain the facts. Winslow's observation therefore raises questions that are pressing today because there are modern counterparts of miasmas, vaguely defined factors that are offered as causes of disease.

Why have experts in health care been slow to recognize infectious causation? Why do they still tend to dismiss hypotheses of infectious causation out of hand and then only gradually warm up to them? My physician, for example, seems like a particularly thoughtful and conscientious practitioner of healing and disease prevention. He keeps abreast of current thinking and tries to keep up with the literature.

When I talked with him in 1998 about infectious causation of chronic diseases, he seemed interested. He agreed that peptic ulcers were caused by infection and thought that infectious causation of atherosclerosis was reasonable and perhaps even likely. But when I pushed the envelope just a bit further, to possible infectious causation of breast cancer, prostate cancer, and schizophrenia, that was too much. He stopped his probing, moved backward, his body language mirroring his intellectual response, gave a not unpleasant but self-assured chuckle, and said, "Now, that's going too far!" I went on making a pest of myself discussing the evidence, but without any discernible impact.

The answer to Winslow's puzzle seems to stem more from human nature than from human knowledge. Humans generate their biases from their everyday experiences and need to struggle continually to recognize them. Physicians may need to struggle even more than the rest of us because they are trusted as the experts in health care, because they are expected to have the answers, because the systems they work with are very complex, and because ethical considerations restrict the normal options for definitive resolutions of uncertainties.

What is it that creates a bias against infectious causation, especially among such highly trained people? Physicians, like everyone else, base their opinions on their experience. Before the microscope was trained on human pathogens, their experience argued against some solid organism causing disease by moving from person to person. Had they been able to see the organisms, the ideas of infection and contagion would undoubtedly have been accepted almost immediately. After all, physicians had been accepting the idea that worms cause disease for more than two thousand years. They could see the worms with their unaided eyes. Mesopotamian physicians could extract the worms that cause dracunculiasis manually, by cutting through the skin and twisting the worms around sticks; and ancient Egyptian physicians purged intestinal worms from patients using drugs. After the worms were removed the patients got better.

Though the physician's eyes could not see any organisms moving from one person to another, the physician's nose could sense that something vaporous and offensive was often released from patients and often wafted into the air from places where diseases were particularly prevalent. The smell itself was sometimes enough to make a person feel ill. Moreover, physicians had some good evidence for miasma-like diseases. They sometimes saw patients who had succumbed to volatile chemicals.

The microscope generated widespread acceptance of infectious causation of most acute infectious diseases because by extending the visual sense, it allowed physicians to see the small organisms and associate them with the diseases they saw in everyday practice. The research and ideas of Pasteur, Koch, and their fellow microbe hunters put microscopes in hospitals and physicians' offices. With microscope eyeglasses, the daily experiences of physicians reinforced rather than weakened many hypotheses of infectious causation, particularly for acute infectious diseases.

Chronic diseases are different. Even with microscopes, the daily experiences of physicians do not reinforce infectious causation of chronic diseases unless the chronic disease develops from an acute infectious disease, as with syphilis and tuberculosis. In fact, daily experience must have done, and must still be doing, just the opposite. A physician does not contract cancer, stroke, dementia, or a heart attack soon after visiting a patient with one of these diseases.

By relying on surrogates, some physicians avoid the bias against accepting infectious causation. For instance, antibiotic treatments can sometimes aid vision much the way microscopes aid vision. The physicians in New York City hospitals who were treating and curing peptic ulcer patients with antibiotics during the late 1940s were using the antibiotics to glimpse infectious causation. The daily experience of those physicians accorded with infectious causation of peptic ulcers because their antibiotic-aided eyes could see the infectious causation of ulcers more clearly than the eyes of physicians who had not been using antibiotics to treat ulcer patients.

Physicians are still considered society's experts on human diseases, and if we want an opinion in the absence of hard evidence, we seek an expert opinion. Winslow would not disagree on that matter. But in contrast with Winslow's interpretation, history does not suggest that physicians are slow to accept infectious causation because they know too much; rather, what they know has been biased against the idea of infectious causation.

Winslow's particular opinions are not widely circulated. Aside from some medical historians, hardly anyone even remembers Charles-Edward Amory Winslow, Yale professor of medicine. The importance of the observation quoted here lies in its persistence. Winslow made his generalization about the slowness of physicians' acceptance of infectious causation as a way of explaining attempts during the nineteenth century to understand acute diseases. With our clarity of hindsight, we can see that the bias against infectious causation has never been eliminated. It still masquerades as expertise.

The causation of cancer provides an eye-opening illustration. Infectious causation of cancer was demonstrated by Francis Peyton Rous, who in 1910 published his discovery that an infectious agent can cause cancer in chickens. Rous, in his 30s, received flak from experts who *knew* that cancers were not infectious. The criticism led Rous, in his 40s, to shift his research away from infectious causation of cancer. The 87-year-old Rous, however, received a Nobel Prize in 1966 for the work he had done more than a half century earlier. That recognition gave a boost to the small cadre of researchers who were pursuing the possibility of infectious causation of cancer in labs scattered around the world. When Nixon's War on Cancer generated a flush of funding, research was supported in both the infection camp and the not-infection camp. Not surprisingly, physicians were more heavily represented in the not-infection camp, though there were plenty of exceptions. When the funding dried up, the not-infectionists received the lion's share of what was left, just at the time when the infection camp was poised to begin a long series of successes.

From then until now, the not-infection camp has been unscientifically dismissing the infectious causation of cancer. With each new bit of evidence, they give up a bit of their argument, adamantly insisting on its general truth. In 1975 they granted that infectious agents cause cancer in animals, but almost never in humans. In 1985 they granted that infectious agents cause cancer in a tiny proportion of human cancers, but such cancers were portrayed as the rare exception. In 1995 they granted that infectious agents may cause about 15 percent of human cancers, but the vast majority of human cancers were caused by something else. In 2005? We can expect the percentage to have risen still higher. There is no sign of a slowdown in the discovery of infectious causation of cancer.

As for the "something else," the not-infectionists point to evidence for oncogenes and noninfectious environmental agents, such as radiation and carcinogenic chemicals. That is where the lack of scientific rigor enters. In rigorous science, rejection of hypotheses requires evidence to justify such rejection. Evidence for oncogenes and noninfectious environmental agents does not rule out infectious causation, because virtually all examples of infectious causation also involve variation in genetic susceptibilities and noninfectious environmental influences. Certainly with regard to cancer, infectious causation works in concert with mutations caused by environmental carcinogens. Hypotheses of infectious causation were and are still victims of a tendency to categorize rather than integrate. Infectious causes of cancer work in the context of genetic and noninfectious environmental causes, not in isolation from them.

During the last quarter of the twentieth century, researchers who made discoveries relevant to the genetic causes of cancer have received Nobel Prizes within several years after the publication of their research.

I predict that historians in the middle of the twenty-first century will praise those who led the research effort into the infectious causes of cancer during this period at least as much as those who were contributing to our understanding of genetic causes. Harald zur Hausen, who has helped decipher the infectious causation of human cancer, will be praised as much as Nobel laureate Harold Varmus, who helped decipher the genetic causation of cancer. It has been about two decades since zur Hausen guided the health sciences toward an understanding of the infectious causes of cervical cancer. Using Peyton Rous as a model, one bit of advice can be offered to zur Hausen and the other pioneers of infectious causation of chronic disease, while waiting for their well-deserved call from Stockholm: Live a long life!

MENTAL MIASMAS

Koch, Pasteur, and the other microbe hunters of the nineteenth century buried the miasma theory by providing hard evidence for the germ theory of disease. So what are the "modern miasmas"?

Consider mental illnesses. Ask medical experts what causes schizophrenia or bipolar disorder and a vague answer will probably be offered, something about biochemical imbalances in the brain. Ask what causes the imbalances and another vague answer will be likely, something about genetic predispositions, stress, and traumatic episodes during critical periods of emotional development. You might even get the vaguest, safest, and least informative pronouncement of all: a mix of environmental and genetic factors. This response tells us nothing, because every disease is a mix of environmental and genetic factors. Infection itself can be considered an environmental factor. An informative answer would identify the particular factors and their specific roles in the process. Today's vague references to something arising from the environment and inherent resistances or vulnerabilities to that something are akin to the statements made by the miasmatists during the nineteenth century. The mind is complicated, but we can pursue a more tangible understanding of mental illness.

Contrast the vague causal explanations of mental illnesses with the view that has been coming from researchers on infectious causation of schizophrenia and bipolar disorder. Evidence of infection with Borna disease virus has been found in about 60 percent of patients with these two diseases by research groups in Japan and Germany. Similar evidence of infection has been found in less than 5 percent of subjects who do not have either disease. The virus infects the brains of a wide range of mammal species, such as horses, sheep, cats, and rodents, and tends to infect those parts of the brain that are known to influence mood.

Experimental infection supports a causal role for Borna disease virus in these two mental illnesses. Infection of rats causes them to lose control of their fear response. Uninfected rats pause after a loud noise, check out the situation, and then resume their activities. Infected rats do not pause. After the sound they seem compelled to keep moving in spite of the potentially dangerous disruption. Normal rats gradually seek out shadowy areas if they find themselves in bright light. The infected rats don't show this potentially lifesaving preference for areas in which they would be less conspicuous to predators.

Research led by schizophrenia expert Fuller Torrey incriminates the protozoan *Toxoplasma gondii,* a distant relative of the protozoan that causes malaria. *T. gondii* has its sexual phase in cats much as the malaria protozoans have their sexual phase in mosquitoes. Cats acquire *T. gondii* by eating other vertebrates infected with the protozoan much as mosquitoes acquire the malaria protozoan by drinking the blood of vertebrates. *T. gondii* stays put inside the cat intestines, where it is unlikely to do much harm. It ranges widely in the bodies of other vertebrate hosts, forming cysts in the brain and other tissues. If *T. gondii* makes a mouse a bit mentally disoriented, all the better for capture by the cat. When the protozoan infects humans from cat feces, it is in its mouse phase rather than its cat phase. Perhaps it affects our minds as though we were the mouse in a mouse-cat cycle of transmission. Whatever the actual explanation, Torrey has found *T. gondii* more commonly in schizophrenia patients than in controls, and has found schizophrenia less commonly where cats are not kept as pets.

Schizophrenia is also more common among people who were born in late winter and early spring, and in urban rather than rural environments. These associations suggest infectious causation though they do not implicate a particular pathogen—cats could play a role in transmitting Borna disease virus or *T. gondii.* Both pathogens commonly infect cats. Knowing that *T. gondii* is transmitted to humans from cat feces, I find myself casting suspicious glances at our litter box, which our cat uses eagerly during the cold months of January through March though rarely during the rest of the year. When I hear the telltale scratching from inside the box, I wonder what the cat might be tracking around the house after it leaves the box. The transmission of *T. gondii* from mother to fetus is well documented; its effects on miscarriages are sufficiently well accepted that pregnant women are advised not to have cats as pets. Perhaps schizophrenia will prove to be a less conspicuous facet of the same maternal transmission, particularly during the winter months, when transmission from cat to mother and baby is heightened by the tendency of all to spend more time indoors.

As mentioned earlier, in the context of atherosclerosis, the fact that two or more pathogens are implicated as causes of a given mental illness is no reason to doubt a role for either one. Eventually we may speak of viral schizophrenia or toxoplasmic schizophrenia in the same way that we now speak of viral pneumonia or streptococcal pneumonia.

Testing infectious causation of mental illnesses is difficult because mental illnesses are defined according to human minds. We feel relatively confident that different people are having similar feelings, partly because we can discuss our feelings with each other and see if they correspond. That confidence begins to disintegrate when we look at behavioral changes in other animals, partly because other animals cannot express themselves to humans as articulately as most humans can express themselves to each other, but also because human brains are more alike anatomically and physiologically than they are like animal brains. How does one tell whether a rodent is paranoid, hallucinating, or depressed? Animal models will not be as definitive for studying mental illnesses as they have been for other illnesses, and ethical considerations preclude human experimentation.

We are left with scientific paralysis if we doggedly adhere to the standards of proof that have been established for acute infectious diseases. To converge on the most reasonable explanation, we will have to change the standards of proof that are required for acceptance of infectious causation of mental illnesses. This conclusion may sound like abandonment of rigor, but actually it is essential to maintain scientific rigor. How many controlled experiments have demonstrated that humans evolved from ancestral apes? None. We accept this explanation because no other explanation we can think of accounts for the entire body of evidence nearly as well. If we did not do so, science would be less informative. We must be ready to use such standards for the causation of mental illnesses and other chronic diseases.

The gold standard of evidence has not been met for any of the chronic diseases that have recently been accepted as having infectious origins. In fact, the abandonment of the highest standards of proof is the very reason there has been a resumption of recognition of infectious causation. No one has demonstrated that cervical cancer is caused by human papillomaviruses or that liver cancer is caused by hepatitis C viruses according to the same standard met by researchers in demonstrating rhinoviruses to be a cause of the common cold. Explanations that do not incorporate infection as a primary cause simply fail to accord with the entire set of observations.

How could invisible specks of protein and genetic material generate something that seems as complicated and human as clinical depression or schizophrenia? All the symptoms that are seen in patients who are ill are characteristics that would make sense if they were present in other contexts and in moderation. Feeling moderately depressed often occurs in response to failure to achieve goals that were too high. Its unpleasant sensations may be a useful human adaptation that causes individuals to reassess a situation and set more realistic goals, much as the pain felt for hours after a burn may cause a person to be more careful around hot objects. Seeing things and hearing voices is great, as long as the things and voices are really there. Infectious agents may simply cause certain circuits to fire when they would ordinarily not fire, or to fire more intensely than is warranted in a particular situation. This kind of effect is easy to envision if one thinks of the brain as a set of complex circuits and feedback loops. A circuit may generate a particular sensation; collections of circuits may be orchestrated into functional modules that generate more complex thoughts and mental skills. If an infection modifies one or more of the feedback loops by destroying some neurons, the sensation or thought process may take on characteristics that are out of place for the environment. Images may be seen that are not there. Sounds may be imagined. Emotional states may be more intense and incapacitating than would be appropriate for the situation.

This kind of logic has been applied to juvenile-onset obsessive-compulsive disorder. Afflicted kids may show a constant fear of contamination by germs, an incessant need to wash their hands, and uncontrollable tics. Brain scans show a swelling in the basal ganglia, a part of the brain that may exert control over such activities. Case studies during the 1990s revealed an association with streptococcal infection, much as streptococcal infections were linked to rheumatic fever during the first half of the twentieth century. Inflammatory responses to the streptococcal infection may short-circuit neurological mechanisms that inhibit normal activities, or they may directly activate the circuits for these activities. The net result is that a person continues to think and do things beyond the point that would be beneficial in a given situation.

Why did people accept stress or stomach acidity as a cause of peptic ulcers? There was no hard evidence demonstrating a causal link. Experts made an appealing argument and other experts accepted it, and then the argument metamorphosed into established "truth" as it was handed down from one expert to another. Why did people believe Freudian hypotheses about mental illness? The same explanation applies. Freud and his followers presented their ideas as something more than untested hypotheses, and that characterization congealed. Although the evidence for infectious causation of mental illnesses is now much stronger than the evidence for Freudian hypotheses ever was, experts tend to give Freudian hypotheses substantial leeway while often dismissing infectious causation.

This knee-jerk dismissal of infectious causation of mental illness is surprising when viewed in historical context. Syphilitic insanity was recognized as a consequence of infection during the first decade of the twentieth century. As was the case with the infectious causation of peptic ulcers and atherosclerosis, attention to infectious causation of other mental illnesses virtually vanished in mid-century. In its place were vague, miasmatic statements about unresolved emotional conflicts, the stress of modern life, familial histories, and genetic predispositions.

Historians will look at twentieth-century explanations of peptic ulcers, atherosclerosis, and schizophrenia and see vague references to lifestyle, stress, genetic predispositions, diet, and upbringing. The miasmas of nineteenth-century medicine were similarly vague references to things that came from a person and a person's environment. Modern miasmas are vague, sloshy causes that are proposed to the exclusion of specific testable causes. Pathogens are specific and testable. A defective gene is specific. Genetic vulnerability to infectious damage is specific. Genes and environment are vague pseudo explanations that are no better than the miasmas of the nineteenth century.

CHRONIC FATIGUE

What could be more vague and clouded than our understanding of chronic fatigue syndrome? The February 18, 2000, issue of the journal *Science* had in its News of the Week section a headline that read, CONTROVERSY CLAIMS CDC LAB CHIEF. The controversy centered on a dedicated director of the division of viral and rickettsial diseases, Brian Mahy. Mahy had been removed from his post for diverting funds. He was not diverting funds for his personal benefit or that of his family but rather to expedite the solution to problems that he judged to be the most crucial. Most scientists working at the Centers for Disease Control and Prevention (CDC) and in infection control would probably side with Mahy. If the directors at the CDC do not have discretion to allocate funds quickly to newly arising problems, transient opportunities for effective disease control may be missed.

This case raises an interesting problem that goes beyond the normal pros and cons assessed by well-intentioned officials. One of the congressionally earmarked targets from which funds were reallocated was chronic fatigue syndrome. That was a dangerous area from which to be reallocating funds, the danger being measurable in units called votes. A large number of voters think they suffer from chronic fatigue or that a friend or loved one does. Funds allocated to chronic fatigue syndrome are like votes in the ballot box. Skeptical health scientists see this as a conflict of interest between disease prevention and politics and feel that valuable money is being squandered. The concern is largely understandable, but in this case Congress may have been largely right and the skeptical scientists largely wrong.

To understand my reasoning, and how I could come to a conclusion that is probably at odds with the conclusion reached by most of the experts at the CDC, we must consider the CDC in historical context. CDC scientists are intellectual descendants of the great microbe hunters and epidemiologists of the nineteenth century: Pasteur, Koch, Snow, and Semmelweis. These modern microbe hunters thrive on being at the front line, solving clearly recognizable problems: Legionnaires' disease in Philadelphia; Ebola in Central Africa; West Nile virus in New Jersey; *Cyclospora,* which arrived in the United States as a stowaway on raspberries from Guatemala; and new, ominous influenza strains anywhere on earth. These researchers are carrying on the honorable tradition, but the microbe hunters of the past already figured out the causes of the major acute infectious diseases. They did so decades before the current crop of CDC scientists were in grade school. Any modern hunter of the microbes that cause acute infectious diseases has to choose from the leftovers. The earlier generations of microbe hunters wouldn't have bothered with Ebola or Legionnaires' disease. They had bigger problems to solve, such as typhoid fever, typhus, yellow fever, tuberculosis, diphtheria, pneumonia, syphilis, and smallpox. A modern hunter of acute infectious disease microbes has to choose between small problems in rich countries or big problems in poor countries, but big problems in poor countries are big not because there is something mysterious to solve. Rather they are big because the infrastructure and poverty of poor countries inhibit the application of the known solutions, such as clean water, vector-proof housing, and vaccination. There the successful interventionist will be not so much a microbe hunter as a strategist who can work within the difficult social, political, and economic constraints that paralyze disease prevention in poor countries.

There are some big infectious disease control problems in rich and poor countries alike. A few are widely recognized infectious diseases that are still around because they are so intractable—AIDS and influenza, for example. But, as discussed, most of the big problems are not yet recognized as big *infectious* problems, and some of these will turn out to be readily controlled when we know enough about them and their vulnerabilities. But these big problems are often difficult to study and easy to dismiss.

Enter chronic fatigue syndrome. It may turn out that most cases of chronic fatigue syndrome are caused by infection and that the infection will be preventable—for example, with a vaccine. If we add up all the people who may suffer from the syndrome in a country such as the United States and multiply that total by the benefits that would accrue if the syndrome were eliminated, then it is quite possible that this total benefit would make the benefit associated with studying and controlling Ebola minuscule by comparison.

Funding of research into the infectious causation of chronic diseases has been inadequate to resolve the ambiguities because officials who are in control of funds have not seen fit to allocate them to the rigorous study of this problem. Funding research into chronic fatigue syndrome may be much more important than funding the study of those acute infectious diseases that grab the attention of CDC personnel because they are more immediate, circumscribable, and preventable over the short run. Invoking the familiar war metaphor, the localized outbreaks of acute infectious diseases are like the Gulf War with Iraq, whereas the more widespread chronic infectious diseases are more like the cold war with the Soviet Union. Action in the Gulf War was more manageable and decisive, but the resolution of the cold war was more important. It is useful to confront problems like Ebola, which are readily controlled with decisive action, but such activities should not distract us from the bigger problems at hand.

There may be many different causes of the diffuse collection of ailments that are grouped together under the label chronic fatigue syndrome. A possible infectious cause is the Borna disease virus. Japanese researchers found it in one third of their patients diagnosed with chronic fatigue syndrome at about ten times the rate at which it was found among healthy controls. This is the same virus that has been implicated in schizophrenia and bipolar disorder.

If the current indications pan out, and Borna disease virus proves to be causally involved in chronic fatigue syndrome, schizophrenia, and bipolar disorder, then reallocation of funds away from chronic fatigue syndrome is another lost opportunity, another case like peptic ulcers, dismissed because those who made the decisions were unaware of the scope of infectious causation. This collection of diseases not only touches the lives of most people, but jars them. The recognition that these diseases are manifestations of the same infectious process could be a step of far greater historical importance than the study of another acute infectious disease that is bound to fizzle out on its own.

Of course, that is comparing apples and oranges. There is no objective formula for comparing a few deaths from Ebola with the millions of people who may suffer from chronic fatigue syndrome, but this lack is no excuse for failing to address the problem. Policy makers inevitably face this kind of dilemma whenever they allocate money to different needs. Whatever the difficulties faced, the analysis is improved by improving our understanding of the apples and oranges at hand. The CDC is the premier organization for controlling infectious disease threats. But if the decision makers there do not have a clear idea of the scope of infectious diseases, then we will be understanding infectious apples but misunderstanding infectious oranges. Given such misunderstandings, we cannot expect to allocate money effectively and control infectious disease threats. The congressional earmarking of funds for chronic fatigue syndrome was an opportunity to resolve this question, and, in that regard, it was a lost opportunity.

The record of the last quarter century of decision making can be used to assess whether the lack of concern over chronic fatigue syndrome is part of a broader pattern. This record does not engender confidence that current decisions are identifying the right targets. We can now see that liver cancer, cervical cancer, infertility, and peptic ulcers would have been good targets of studies of infectious causation and epidemiological spread at any time during the twentieth century. Such studies, if they were even considered, were given very low priority. We need to be asking whether our policy makers are continuing to make the same mistakes.

Do any of the chronic diseases that are not now accepted as being caused by infection actually have infectious causes, and can such infectious causes be controlled? These are major challenges for those who hunt and study microbes in the twenty-first century, challenges that, if met, may have as much positive effect on human health as the challenges that were met by those who applied the germ theory to acute infectious diseases. But the challenges are no longer the easy-to-recognize acute infectious diseases. Rather they are the stealth infections that insidiously generate a cancer or neurological damage or a lesion in an artery wall. These diseases develop so gradually that they may be dismissed as the natural wear and tear of life rather than infectious disease. They can be as lethal as many of the acute infectious diseases that were solved by the microbe hunters of the late nineteenth and early twentieth centuries, but they do not kill in a way that gets the adrenaline rushing, and they do not kill as a result of obvious infection. What is a modern-day microbe hunter to do? Investigate the chronic diseases that most experts do not think are caused by infection, and figure out how to control them.

PART III

BEYOND THE FEAR OF INFECTION

Managing Microbes

CHAPTER ELEVEN

Reverberations Across Society

As we see the scope of infectious causation broadening, not just medicine but society as a whole will change. Though it is hard to gauge the extent of the changes, it is clear that society was changed dramatically by the first round of the germ theory. The second round, which will reveal the scope of chronic infectious diseases, may have even more far-reaching effects. It may challenge our notion of our vulnerabilities, our identities, and our potentials. By the time all the feedback loops are accounted for, people will look at themselves and the human condition differently. Few lives will remain unchanged; few enterprises will continue as they had before. This scope ranges from the subtle to the startling, from the immediate to the circuitous.

BYTES FROM DISEASES

The decisions of microbe hunters about what to look for is not the only thing that could foster the second round of the germ theory. The infrastructure of scientific information gathering could be altered as well. In particular, anecdotal information could be used more effectively than it has in the past.

The CDC and the FDA jointly administer a program called VAERS, which stands for Vaccine Adverse Event Reporting System. VAERS accepts reports from physicians, patients, and family members who have observed problems that might be attributable to the administration of vaccines. VAERS is a good idea—it gives the health sciences antennae to sense problems far earlier than if recognition depended on some insightful person somewhere mustering the time and effort to make a case for adverse effects of vaccines. With the current standard of vaccine safety, one physician may never witness enough adverse effects to get a sense of whether they are truly effects of the vaccine or simply coincidences, unimportant anecdotal observations. If the people receiving the flood of anecdotal information are astute and insightful, they will see associations that warrant detailed, carefully controlled studies.

But VAERS does not go far enough. A parallel reporting system is needed for detection of positive effects of treatments on chronic diseases. Let's call it EARS, for Effects of Antimicrobials Reporting System. I have received a great deal of mail from readers describing anecdotal associations between the use of a drug and improvement in a chronic disease. Many of these associations may be spurious, but one cannot tell one way or the other from anecdotal reports, just as one cannot tell whether an adverse event that occurred after a vaccine was an effect of the vaccine or a coincidence. If hundreds of thousands of such reports were accumulated in one database, the real effects of a drug could leave behind telltale statistical associations that could provide the basis for controlled epidemiological and clinical studies to determine whether, in fact, it does have a positive effect. If so, the logical next step would be to investigate whether the positive effect results from an antimicrobial action.

If such a system had been put in place in the 1940s, antibiotic cures of ulcers could easily have become part of mainstream medical practice by 1955. Some important treatments of chronic disease have been recognized by just such anecdotal observations. The effectiveness of the antimalarial drug Plaquenil (hydroxy chloroquine sulfate) against the autoimmune disease lupus was revealed anecdotally when a lupus patient improved after taking Plaquenil prophylactically during a visit to a malaria-ridden area. The drug also works against rheumatoid arthritis. Does Plaquenil help ameliorate these conditions by an antimicrobial action? No one seems to know.

Sometimes insight may come from the reciprocal association: a drug used for a chronic disease shows effectiveness against a known infectious disease, thus implicating infection in the chronic disease. Lithium treatment of schizophrenia and depression offers an illustration: when patients were treated with lithium, their herpes infections improved. This association is one more piece of evidence implicating viral causation of these mental illnesses.

The historical record reveals that many such associations would probably be found. Any one of them might fail to turn up an infectious cause. The beneficial effects of Plaquenil on lupus and rheumatoid arthritis, for example, might result from the suppression of another pathogen that is triggering a malfunction in the immune system; or the beneficial effects may result from direct interference with an overreactive immune system. Even if the benefits uncovered by a surveillance program turn out not to result from an antimicrobial effect, the information on their effectiveness against chronic disease is useful. If they do have an antimicrobial effect, then they may point the way to the causative infectious agent. No one knows how many more bits of valuable anecdotal information are being lost because we do not have such a surveillance system in place. If peptic ulcers are any indication, the consequences might translate into additional decades of ineffective treatment and many thousands of people needlessly suffering and dying from chronic diseases with infectious causes.

THROWING THE BOOK

The legal profession will not be the same. Currently, for instance, actions are taken against polluters when clusters of disease are linked to the release of disease-causing compounds. Consider the case of Keil

Chemical in Lake County, Indiana. In 1991 records indicated that Keil Chemical released about nine hundred tons of a potentially carcinogenic solvent called ethylene dichloride; its city permit allowed the company to release 0.02 ton. From 1987 to 1998, the cancer incidence in Lake County was about 20 percent above the incidence that would have been expected based on national averages.

The conventional perspective is that cancer is generally caused by genetic alterations by physical agents. According to this perspective, the evidence looks incriminating, even though experts warn against jumping to conclusions. In the Lake County case, an epidemiologist from the National Cancer Institute warned that "we don't understand what causes ninety to ninety-five percent of childhood cancer." But these warnings seem like overly cautious science-speak to people who want justice rather than statistics. When scientific proof of causation is not available, the affected people and the legal professionals base their actions on whatever evidence is available. Disease does not wait for science. If there is an excess of cancer deaths and an excess of a potentially carcinogenic chemical, cause and effect will be inferred unless some viable alternative is presented. If a person hears a shot from a room, walks into the room and sees one person dead on the floor from a bullet wound and another person holding a smoking gun, that too is just an association. Unless the person holding the gun has a good alternative explanation handy, the legal profession will tend to assume that the association reflects causation.

In the absence of knowledge about the spectrum of disease causation, experts may unwittingly fan the flames. In the Lake County case, physicians apparently told the parents of two daughters afflicted by rare and different cancers that having both cancers in the same family would be extremely unlikely under normal conditions. Presumably that conclusion was derived by multiplying two small probabilities together to obtain the extremely unlikely joint occurrence within one family. The validity of that conclusion depends on the cancers and what causes them. If the rare cancers were both caused by a common pathogen, then having two in a single family might not be so unusual. Many infectious diseases co-occur frequently within families because family members are more likely to be exposed to the same infectious agents than are two people drawn at random from the population. Medical science has only recently begun to understand infectious causation in the common, well-studied cancers, such as liver, stomach, and cervical cancer. Almost nothing is known about the causes of most rare cancers, certainly not enough to conclude that you can multiply the two small probabilities of having either cancer in a family to obtain the probability of having both in the same family.

The new understanding of infectious causation dramatically changes this kind of analysis by drawing attention to feasible alternative explanations for the clustering of cancers and other chronic diseases. Cases of infectious diseases typically cluster either because the potential for transmission varies from place to place or because infectious outbreaks spread from somewhere. Clusters of cancer would therefore be expected for cancers that are caused by infection. If such a cluster happened to occur near the release of a carcinogenic chemical, then the chemical pollutant might be falsely accepted as the cause of the cancer. But infectious causation makes the morass even more ambiguous. If no clustering is associated with the release, the oncogenic chemical could be falsely exonerated. To see this point, imagine that the release did cause some of the cancer. Imagine further that the cancer rates in the vicinity would otherwise have been relatively low because an infectious cause of cancer happened to be rare in the area. An observer might not see any increased incidence of cancer in the area because the lower background rate might mask the effect of the chemical carcinogen. Joint causation further complicates the situation. Noninfectious causes of cancer may exacerbate infectious causes.

Before we throw the book at alleged perpetrators, we have to look broadly to see the right targets. Historically, the EPA and CDC have been set up to focus on different targets: the EPA has focused on noninfectious causes of disease, and the CDC on infectious causes. This distinction may become increasingly anachronistic and unproductive as we find more disease that is attributable to both causes.

GROWING OLD

What is growing old, anyhow? A half century ago, people thought that the most obvious aspects of growing old—senility, strokes, heart disease, and cancers—were part of the natural process of aging. Now we recognize that they may often be wreckage from our collisions with the microbial world. If most microbes deal with us benignly, then we are compelled to ask how much of the less obvious part of aging is caused by microbial fender benders. Logic tells us that it may be much. If so, what can we expect from human life simply by preventing the damage from our encounters with microbes?

The diversity of activity among the elderly gives us a clue. The bodies of some 50-year-olds are falling apart, whereas some people pushing 80 seem to be cavorting like teenagers. I know this from personal experience because my mother, Sara Jeanne Ewald, is one of these teenagers in her late 70s. She does not respond in ways typical of a person of this age group. For instance, she was run over by a truck in November 1998 and brought to the hospital with a badly fractured pelvis, a broken set of ribs, and a punctured lung. She left the hospital in December and was walking with a cane in January. In April she discarded the cane and departed on a European tour with her boyfriend, leaving in her wake doctors and nurses who were happily scratching their heads in disbelief.

Could the recovery rate from such injuries depend on whether someone was lucky enough to be resistant to chronic infections? I don't know. Many of the elderly and some middle-aged people have problems with osteoporosis; if infections play a role in this condition, then an elderly person who is resistant to such infections might be especially well able to heal broken bones. Sara Jeanne has been incredibly resistant to acute infectious diseases throughout her eight decades of life. "I must have a strong immune system," she would often say to me as I was growing up. Perhaps that may help explain why she is zipping around like a college student on spring break instead of being hobbled by the ailments of old age.

Of course this account is just an anecdote. But what is an anecdote? In an effort to be scientifically rigorous, twentieth-century medical science has made *anecdote* a dirty word. Ardent attempts to codify rigor have stripped us of the benefits anecdotes provide. Anecdotal observations are essential for rigorous science because they provide possible clues to the solution of medical puzzles. Their true value often cannot be discerned without follow-up studies. They may turn out to be junk or gems. When anecdotal observations are followed up with careful studies, some will be recognized as spurious coincidences, whereas others will be recognized as the signposts that guided research to new breakthroughs.

The vision of medicine is sometimes blinded by the average. Any large cohort of 80-year-olds will include some who are youthfully active and others who have become immobilized by the "process of aging." We see the same in 70-year-olds and 60-year-olds. But as the cohort becomes younger, our sense of what is normal changes. We begin to see the debilitation as something out of the ordinary and therefore deserving of a special explanation. We therefore begin thinking of the debilitation as disease rather than as part of the normal process of aging. Once this transition in thinking occurs, we are spurred to understand the cause of the illness. Perhaps when we understand the full scope of infectious causation and effectively prevent its damaging outcomes, vibrant 80-year-olds will be the rule rather than the exception.

ARTISTIC GENIUS

If mental illnesses are caused by pathogens, what do we make of the associations between mental illnesses and pioneers of art? Depressive illnesses, for example, are more common in novelists, painters, playwrights, and poets than in the general population. If these illnesses are caused by infection, is it appropriate to conclude that some aspects of artistic talent are attributable to infection? Did Van Gogh owe his artistic genius as much to a pathogen as to his genes or upbringing?

One might argue that the causation goes the other way around, that artistic lifestyle predisposes artists to infections like the Borna disease virus. But schizophrenia, at least, develops more frequently in people who are born in late winter, suggesting not only an infection, but an infection early in life. It is difficult to believe that one's future career will predispose one to infection as a baby. Perhaps parents might become infected if they had artistic leanings; they might then pass on both the artistic predilection and the pathogen to their offspring. Though possible, this kind of rescue seems reminiscent of the epicycles that were needed to rescue the geocentric theory of the universe.

This linkage between infection, mental illness, and creativity may seem premature, considering that psychiatrists still disagree over the boundaries between schizophrenia and depressive illnesses, and even whether there *is* a boundary, as opposed to an indistinct merging of each into the other. Sometimes understanding the causation of illnesses resolves confusion, as it has for the different kinds of hepatitis and pneumonia. The associations between bipolar disorder and Borna disease virus and between schizophrenia and Borna disease virus, for example, raise the possibility that parts of each of these two illnesses may be manifestations of a particular infectious process, much as rheumatic fever and sore throats may be two manifestations of a streptococcal infection. Finding a common infectious cause could help resolve some of the current disagreements, which are based on alternative groupings of disease symptoms. Common infectious causes might also reveal which, if any, infectious agents are responsible for the creative aspect of mental illnesses and perhaps suggest ways to modify infections to preferentially reduce the damaging aspects of the conditions.

Symptoms like hallucinations need not be regarded as a cause of artistic achievement, any more than an athlete's injury is assumed to be a cause of athletic achievement. But getting injured is almost always part of an athlete's experience. Having said this, however, it is important to add that injury can sometimes contribute to recognition of someone as a great athlete—an athlete who has attained victory after coping with a major injury is generally viewed as having greater talent than one who has attained the same victory without having to overcome injury. Our knowledge of Van Gogh's mental torment makes his paintings all the more arresting. Injuries may contribute something directly to athletic achievement as well, by building a sense of character—the athlete who has successfully overcome one difficulty may be better prepared to overcome a greater one. A tendency to hallucinate or be depressed might contribute in some way directly to creativity, but the argument for infectious causation of artistic creativity does not require it. It requires only that some effect of infection has allowed the artist's mind to deviate from the norm, and has thereby contributed to the artist's achievement, just as the athlete's genetic makeup and training contributed to the athlete's achievement. The interplay between art and science may take on new dimensions as our new century progresses.

CHAPTER TWELVE

Biobombs

When I bring visitors to the Amherst College campus, I usually put them up at the Lord Jeffery Inn. The three-story structure is a prominent member of a row of immodest giants that look down on the Town Commons. Its whitewashed brick exudes New England charm in a self-important Yankee way. It does have some reason to boast. It stands in the middle of a favored hollow. In 1812, Noah Webster moved in at one end of the Commons, a block to the right of where the inn now stands. He was brought there by his stubborn determination to write a dictionary that would help unify a young country through a common language. Realizing that he had to choose between this dream and the worldly but pricey environs of New Haven, Connecticut, he reluctantly moved to the backwater town of Amherst, abandoning mainstream society for, as he put it, "a humble cottage in the country." He joined about twenty-five other families whose life was organized around the Commons, which was then a gentle slope of birch trees on which grazing privileges were shared on a rotating basis by the town's cows.

At the Webster end of the Commons, just twenty steps from where the inn is now located, stood the one-room schoolhouse that so appalled Webster that he set to raising money to build Amherst Academy almost as soon as he arrived. Within three years the academy had been built a block uphill from the schoolhouse, just beyond the corner of the

Commons. Webster postponed his dictionary by several more years to raise money for a college even more earnestly than he had for the academy, partly because he felt poorly treated by Williams College, which had rejected his proposal to affiliate with Amherst Academy. The episode started a usually friendly rivalry between Williams College and Amherst College, which continues to this day. The first austere buildings of Amherst College still stand at the top of a hill two blocks to the inn's left, at the other end of the Commons from Webster's "humble cottage."

In the year of Webster's death a teenage Emily Dickinson, living a few doors down the road from Webster's farm, could look out her bedroom window toward the Commons. Webster had returned to New Haven by that time, but Dickinson still felt his legacy as she walked alongside the Commons past the Webster farm to attend, somewhat irregularly, the academy. The legacy was also felt by the young poet Robert Frost and the young physicist Niels Bohr, at the other end of the Commons, as they struck up a friendship at the college a century after Webster busily raised money for a school that would link the Arts with the Sciences.

The visitors invariably enjoy their stay at the Lord Jeffery Inn, but it seems as though they should be feeling apprehensive, just as they would if they sat down to eat at a diner called the Typhoid Mary Café. As a visitor pulls up a blanket on a cold winter night at the inn, he might remember that Lord Jeffery Amherst was the guy who is famous for giving smallpox-laden blankets to the Indians of western Pennsylvania.

BIOLOGICAL WARFARE

What Jeffery Amherst actually did or did not do is unclear. Written records suggest that this story may have been based on some advice Amherst gave in his capacity as Britain's commander in chief in North America. In 1763, Fort Pitt, an outpost at the western edge of colonial

Pennsylvania, was in danger of being seized by the local Indians, who had had quite enough of Yankee hospitality. They had already taken over all the nearby outposts and killed the inhabitants. If Fort Pitt were to fall, the Pennsylvania colony would contract to its eastern core around Philadelphia. Amherst sent a letter from his base in New York to the ranking officer for western Pennsylvania, Colonel Henry Bouquet, who was based in Philadelphia: "Could it not be Contrived to Send the *Small Pox* among those Disaffected Tribes of Indians? We must, on this occasion, Use Every Strategem in our power to Reduce them."

Bouquet wrote back, "I will try to inoculate the ——— with Some Blankets that may fall in their Hands, and take care not to get the disease myself." Bouquet apparently deleted the name of the Indian group to protect himself or the plan.

The response from Amherst makes his opinion clear: "You will Do well to try to Inoculate the Indians, by means of Blankets, as well as to Try Every other Methode, that can Serve to Extirpate this execrable Race."

The journal of the captain at Fort Pitt, however, indicates that he had already given smallpox-laden blankets and a handkerchief to the Indians before receiving such orders from Bouquet. The use of smallpox-contaminated fabrics apparently was generally recognized as an option that could be used if other methods failed, and sometimes even if they didn't. Smallpox broke out severely among Indians the following spring, though the outbreak may have resulted from sporadic cases that had been occurring among the Indians even as the captain at Fort Pitt was preparing his blankets. The colonists, having already acquired immunity, were little affected. Fort Pitt was held and eventually grew into a hub of western Pennsylvania: the town of Pittsburgh.

Not quite two hundred years later, after Nazi panzer divisions reached Stalingrad, a strange "German disease" hampered their advance. A surprisingly high incidence of tularemia soon occurred among Soviet civilians and troops in the area. Though the bacterium that causes tularemia is similar to the agent of the black plague, tularemia generally stays put in its normal transmission cycle, which involves small mammals and ticks. Human cases occur sporadically if, for example, a person is bitten by an infected tick or is butchering an infected rabbit. Three decades after the Stalingrad outbreak, a promising young medical student named Ken Alibek was asked by his Soviet professor to analyze the unusual cluster of tularemia cases. After intensive study Alibek reported his conclusion to the professor: the epidemic must have been caused intentionally. The professor then told Alibek to "forget you ever said what you just did. I will forget it too . . . never mention to anyone else what you just told me. Believe me, you'll be doing yourself a favor."

In one sense tularemia was an effective weapon for the Soviet goal. It helped stop the Nazi advance. But as was the case in western Pennsylvania nearly two centuries before, the weapon was crude and unpredictable—it could fizzle or backfire. The sequence of events in Stalingrad probably began with spraying of the Germans when the winds were favorable; the collateral damage to Soviet civilians and troops may have resulted from a change in wind direction or from infection of rodents, which then dispersed freely across the battered landscape.

Learning their lesson from the experience, Soviet strategists shifted their tactics to use of biological agents against targets that were further behind enemy lines. But even this alteration would not have solved the problem. If biological weapons are successful, the enemy territory may soon be the property of those who contaminated it, and a heavily contaminated prize is not very attractive; moreover, the mobility of people in war makes the site of release a poor predictor of the spread of damage.

Our emerging knowledge of disease evolution, considered in the context of the biological weapons used in past wars, raises a weighty question. As we learn more of how evolution creates virulence—indeed, as we learn how to manipulate that process—are we likely to begin constructing a new military technology based on our new knowledge? There are dangers, but there are also reasons to think that the dangers can be lessened.

With just a little bit of thinking, military strategists and militaristic dictators who are tempted by the low cost and destructive capacity of biological agents recognize the general lesson: biological organisms may be scary, cheap, and accessible, but they are poor military weapons. Even when biological weapons could provide a tactical victory, their use opens the door to long-range difficulties. If the stronger side initiates the use of biological weapons for a tactical advantage, it opens the door to their use by the weaker side. Though biological weapons offer little chance of victory for the weaker side, even a weak opponent can cause a great deal of retaliatory damage to troops or civilians. So the more powerful side has an incentive not to open that door. A similar incentive applies to the weaker side, which risks retaliation from the stronger side's conventional weapons as well as any biological weapons the stronger side may have. Escalation to an air war provides a contrast: the side that has air superiority and initiates the use of air attacks may suffer little from the other side—as long as the air superiority holds up.

Clearly there has been a hesitancy to deploy biological weapons. They have been used when a weak opponent has little else to lose from retaliation, as was the case with the Soviets in Stalingrad, or when a weaker opponent does not have biological weapons for retaliation, as was the case in colonial Pennsylvania. In Stalingrad, the Soviets were not risking a great deal of damage from reprisals, because their situation was dire and because the Nazis did not have a strong biological weapons program and were already using everything in their full arsenal. In western Pennsylvania the situation at Fort Pitt was dire, though the situations in Philadelphia and New York were not; but opening the door was not so risky, because the Indians had neither biological weapons nor overwhelming military power kept in reserve for retaliation.

This logic provides an important though somewhat discomforting lesson: a power that wants to avoid being the target of biological weapons during wartime had better have some means for reprisal. It could do that by having biological weapons of its own, or, as the United States has done, by having such an overwhelming reserve capacity for destruction that the use of biological weapons by a less powerful adversary is deterred.

Still, even in such cases biological weapons are weapons of last resort. When a city is being overrun or a fort is under siege, biological weapons may turn the tide, but they may do so at a great price and with little predictability. These problems are inherent among living microorganisms because the reproduction, spread, and modification of living organisms, and the feedback loops among these influences, are too complex to predict and control once the organisms are unleashed.

The modification of disease organisms for use as weapons during wartime has been a frightening threat ever since the use of chemical weapons during World War I made observers ask whether the science of weapons development should be limited. A deeper analysis of the drawbacks of biological weapons assuages these concerns to some extent by revealing that biological agents tend to generate poor weapons for warfare. But what about biological terrorism?

THE THREAT OF BIOTERRORISM

The goals of terrorists differ from the goals of military strategists, so the drawbacks of biological weapons may not be as disadvantageous for the terrorist. Retaliation against shadowy groups may be difficult; the potential for spread from the site of release may not be so costly if the perpetrator's "troops" are not in the vicinity; and unpredictability may not be so disadvantageous if the goal is havoc. When the disadvantages of biological weapons are not so prohibitive, the threat may be greater. Not surprisingly, strategists and security groups in governments tend to be more concerned about the use of biological weapons by terrorists than in warfare.

As with other terrorism, the actual threat from bioterrorism depends on how self-destructive the terrorists are. The more self-destructive they are, the greater the threat. Still, biological weapons have major drawbacks even for terrorists.

Terrorists attempt to attract attention to their causes or gain power by creating spectacular damage. They tend to benefit if a sudden strike is followed by a long lull without any damage. The lull causes people to worry about the horror of what could happen relative to the current peaceful situation. Infectious weapons tend to generate the opposite pattern. The initial attack quickly grows to fearful proportions but then wanes slowly. As the danger lessens during the long waning phase, the fear among the psychological targets is replaced by acceptance of the reality of the situation, a hardening of resolve, and a desire for vengeance. This change in attitude is bad for the terrorists because the hardening of resolve and desire for revenge persist as the frequency of disease diminishes. People do not think well in terms of large numbers, but do respond to the stories of individual people. The ongoing personal interest stories during the waning phase of an infectious outbreak increasingly anger people because as the experience continues, each person has a greater chance of having a direct exposure to someone who has suffered from the disease; each is more likely to have a personal reason to take revenge on the perpetrators. As the time following an infectious terrorist attack increases, the resolve to stamp out the perpetrators may continue to increase.

This pattern of impact is quite different from the pattern that would occur after a discrete terrorist act, such as a bombing. As the time following a bombing increases, the number of severely affected people does not increase, and the anger tends to subside. Yet the fear of a recurrence persists, and it is this fear that transforms the shock of the initial damage into a response that provides terrorists their goal, whether that goal be recognition, power, or appeasement. A bomb is a severe cut that heals, whereas an infectious outbreak is a festering wound. After a wound heals, the desire for revenge decreases, though the fear of a similar wound persists. If a wound festers, the desire for revenge is maintained with the continuing injury. Infectious weapons thus make the terrorist's existence more tenuous by mobilizing persistent opposition to the terrorist group.

But there is more bad news for a terrorist who uses infectious weapons. An infectious weapon saps its own power. The power is gained through fear of a second attack, but an infectious outbreak saps this fear because the actual danger of a repetition of the attack is reduced. The outbreak itself has protected the survivors against a second attack by the same organism through the immunity that the first attack generated among the survivors. This immunity results from both the infections generated in the first attack and the increase in vaccination against the weapon that arises in the wake of attack. If the first attack was a howl, the second would be a bark, and the third just a yip. Instead of demonstrating that the terrorist organization is a force to be reckoned with, the subsequent attacks would be a display of relative impotence.

Of course, the terrorists could switch to another weapon, but then they would have to start over with the process of educating the targets about the danger the new agent poses. Without this education, the threat of a next agent on the list is too abstract. A bombing reported in vivid detail by the news media is a far more tangible threat than the imagined pathology associated with a new infectious agent. And bombings do not generate immunity in the population.

If terrorists played the game of switching infectious agents, they would soon run out of options. Only a handful of agents would be suitable terrorist weapons. If the terrorists chose the most frightening agent to begin with, they would have to shift to progressively less effective agents. Once again they would be confronted with the strategic problem of their howl's diminishing to a yip.

THE CHAOS OF DISEASE

Though the unpredictability of biological weapons may be less important to terrorists than to war strategists, it may still be a major drawback. Even terrorists need a ballpark estimate of the damage they are likely to cause, and the sources of unpredictability are myriad. When terrorists use weapons, they are looking for a psychological effect, and such effects are likely to vary tremendously depending on the amount of damage inflicted. If a microbial attack fizzles, then the terrorist group would look like incompetent bunglers. If it got out of hand, killing, say, a million people instead of a thousand, then the reaction against the terrorist group might be so strong and unified that the terrorist act would be suicidal for the organization.

The scope of damage is particularly difficult to estimate for biological weapons because the rate at which infections spread can increase geometrically. If one infected individual infects ten others, and each of those infects ten others, then a hundred infections will result after two rounds of transmission, a thousand after three rounds, and so on. If such an attack was recognized early enough to curtail the outbreak by only three cycles, the overall number of infected people could be reduced about a thousandfold. Conversely, if people were particularly slow to recognize what was happening, the outbreak could be several orders of magnitude greater. A terrorist therefore has a great chance of causing too little or too much damage depending on whether the outbreak is recognized a few hours earlier or later than planned.

Dosage presents another source of unpredictability. Depending on the dosage, the net effect of microbial pathogens in most infections can vary from an infection without any symptoms to lethality. If most people get low dosages, the net effect of an attack might involve something more akin to vaccination than to a lethal epidemic. Those people who get high dosages and who are particularly vulnerable will tend to have very short incubation times. Variation in incubation times contributes to the uncertainty of the effect of a microbial attack by generating a large potential variation in the time of recognition of an outbreak.

Dosage depends on all sorts of variables that are out of the control of the perpetrator, variables such as humidity, light intensity, and temperature. Most scenarios for the use of infectious weapons involve respiratory tract pathogens, which tend to be very sensitive to ultraviolet radiation. This sensitivity has several important consequences. It means that transmission of infection and effective dosages will be reduced if weather raises the intensity of ultraviolet light, or if weather causes people to stay outside rather than inside. It also means that the degree of within-building transmission relative to between-building transmission will depend on weather. The transmission dynamics of these two types of spread are bound to be different.

Even if terrorists could master all the available knowledge from epidemiology and microbiology, tremendous uncertainty would still be associated with the effects of their actions. The best minds in these areas of science are nowhere close to being able to predict such details with accuracy. Unless terrorists can outclass the best minds in epidemiology, we can safely assume that the use of biological weapons would be a bad move by any terrorist looking to cause a particular severity of infectious outbreak.

If attackers underestimated the scope of the outbreak, they would be likely to cause damage to unintended targets. Infectious agents do not restrict themselves to ethnic or national boundaries. Considering the uncertainties, we can expect consequences that may vary by several orders of magnitude. Predictive errors of much smaller magnitudes could subsume not only the targets of the attack but the attackers' beneficiaries and allies as well, an extreme case of death by friendly fire. Such unintended targets may be in New York City one morning and at any other airport in the world the next day, spreading infection to other unintended targets.

Terrorists could try to vaccinate those whom they wish to protect, but germs evolve. A germ generated during an outbreak might change its makeup, especially if it was a laboratory variant. A terrorist cannot tell whether a microbe will rapidly evolve changes in virulence or transmissibility, nor whether a vaccine that was protective against the germ that went into a target population will work against the germ that comes out.

The terrorist needs sudden and spectacular devastation. Use of infectious agents compromises this goal because the devastation they can impose is constrained by a trade-off: high transmissibility is generally coupled to reduced lethality. If a pathogen is rapidly lethal, then it is generally not very transmissible. Terrorists attempting to cause a highly lethal outbreak will tend to have a short-lived outbreak. But if an infectious weapon is nontransmissible, it loses much of its fearful aspect, which arises from the prospects of a limited release spreading to engulf a large population. If transmissibility is maintained, virulence is reduced. The smallpox virus—probably the most dangerous transmissible infectious weapon—offers an illustration. The virus can be expected to kill about 20 percent of those who are infected. Although the number of deaths caused by the use of this agent would be horrific, it is important to realize that in the wake of a smallpox battle four out of five victims would survive. If a bomb killed only one out of five of the targets, the terrorist action would be considered a failure. More important for the terrorists, the survivors would be angry. The victims would be ready to take revenge on those responsible for killing their family and friends, and world opinion would be on their side. Draconian measures to root out the responsible terrorists would be all the more likely to be considered justified.

BEING PREPARED

Although terrorists would have to be on the duller side of the IQ curve to use biological weapons, it would be foolish for us to believe that such terrorists do not exist. We have been referring to terrorists as though they were fairly homogeneous with regard to their motivation, goals, intelligence, and rationality. We would be wise to prepare for heterogeneity. In this business it would take only one renegade to cause havoc. Such outlier terrorists may indeed use infectious weapons, and in so doing simultaneously demonstrate the clumsiness of infectious weapons and the self-destructive consequences for those who deploy them. It is a lose-lose situation for which we need to be prepared. The better prepared we are, the more clumsy the terrorist will appear.

How can we prepare for an epidemic initiated by terrorists? The most obvious countermeasures are surveillance and interception. Although these strategies could effectively counter many of the threats, believing that they could provide complete protection is probably wishful thinking. Similarly, generating complete vaccination coverage for every possible infectious agent is probably both logistically and ethically unfeasible. Mandatory vaccination against smallpox, for example, would handicap or even kill a tiny percentage of those vaccinated; a tiny percentage of an enormous vaccinated population translates into a large number of people being harmed. Imposing this harm on people for an uncertain benefit would be ethically questionable and would probably be socially divisive over the long term.

Fortunately, complete vaccination is probably not necessary to deter terrorism. Infectious weapons are poor terrorist weapons, and incremental increases in vaccination coverage make the infectious weapons incrementally poorer. Simply making the smallpox vaccine widely available for voluntary use would increase the percentage of people who would be immune to a smallpox terrorist attack. Informing the populace about the true threat of the smallpox virus as a terrorist weapon would increase this percentage further. If half the population chose to become vaccinated, the bang that the terrorist would achieve would be reduced by at least half, and perhaps by much more because with increasing vaccine coverage the infectious spread to the remaining susceptibles would be reduced. Making vaccines available for voluntary usage thus makes a poor terrorist weapon even poorer. The poorer the weapon, the less likely it is to be used. The goal is to make the weapon so poor that even the terrorist with the lowest IQ can see that using it is not in his best interest.

By making vaccines available on a voluntary basis, we would also be developing an infrastructure for vaccine preparation that would speed the response to a terrorist act. It would also vastly reduce the number of additional vaccinations that would be needed in the wake of an attack. If half the population is already vaccinated, the task of eradication of the pathogen would require fewer than half as many additional vaccinations than if no one were vaccinated.

There is a simple reason we have seen frequent bombings and almost no infectious terrorism over the past century in spite of ready access to infectious agents. The terrorists are generally too smart to use them. After thinking about the consequences, they fall back on the tactics that create destruction and fear but leave open avenues for power and recognition to be gained by the terrorist organization in the wake of the damage. Bombing appears to be a preferred method because the human loss that occurs can be explained away as collateral damage, even though it may have been a primary goal. If a building is destroyed, the terrorist can always justify the act as an attack on the inequities of power. The bombing of the World Trade Center could be rationalized as an attack on a symbol of hedonistic capitalism. Such a rationalization may lower the outrage among the have-nots, who are envious of the wealth the building symbolizes. These bystanders may accept on some level the idea that the symbol was the primary target rather than the people. This cover will not work for infectious weapons, because they destroy people but leave the symbolic infrastructure intact.

DESIGNER BUGS

One of the scenarios that surfaces in discussions of biological terrorism is the possibility of a genetically engineered superbug for which there is no defense. The ability of molecular biology to accomplish such an engineering feat is limited, however. A superbug would need to have the right mix of transmissibility and virulence, and the interplay between different genes in the genome is not sufficiently well understood to allow this kind of designer bug to be generated. Nasty variants can certainly be created in laboratories by trial and error, but this possibility existed long before the precision of molecular engineering arrived on the scene.

Animal models won't do the trick either, because repeated transferring of a pathogen between animals of a given species tends to adapt the pathogen to that species and makes it less able to live in and harm humans. One can therefore obtain a highly lethal and transmissible pathogen for mice that is unpredictable in humans. To generate a pathogen that would be both highly damaging and transmissible after being released in a human population, pathogens would need to be transferred among humans in the same way that virulent pathogens can be generated in an animal species—by transferring the pathogen among many individuals of that host species. This option is so abhorrent to most people that we should be able to suppress it. Even if a psychotic individual was willing to do such research, the project would have to be so large, and the probability of being appalled would be so great for those involved, that word would probably get out; when the word did get out, it would most likely be possible to obtain overwhelming approval for drastic measures to end the work and remove from power those who instigated it. The project would be so self-destructive that only the most foolish would engage in it. Still there are self-destructive psychopaths, and there are historical precedents, such as the experiments by the Japanese on prisoners during World War II. The knowledge that superbugs could be created by using humans as experimental subjects should generate multinational public support for efforts to monitor any research activities using human subjects and to place the investigations and dismantling of any such projects above any pleas of national sovereignty.

The potential dangers of such a superbug are illustrated by the history of contact between people during the colonial period. Before this contact, something akin to an experimental passaging of pathogens in humans was carried out. This passaging occurred over thousands of years in European populations, who coevolved resistance to endemic pathogens such as measles, mumps, and smallpox. When these pathogens were introduced into human populations who did not share this coevolutionary history, the lethality of each was ferociously elevated. Whereas smallpox killed one in ten in Europe, it killed most of the New World people it infected. Measles was transformed from a disease that killed about one out of every thousand infected Europeans to a disease that killed New World people the way smallpox killed Europeans. Mumps, which killed less than one in a hundred thousand Europeans, killed New World populations the way measles killed Europeans. The net result was the destruction of New World populations by more than 90 percent. Throughout the arms race between Europeans and pathogens, the pathogens were continually selected to break through human defenses and be transmissible. The evolved and acquired immune defenses of Europeans caused the pathogens to evolve particularly aggressive characteristics that, though held in check in European populations, met with little defense in populations native to the New World.

This natural experiment of human history should keep us from being complacent about the damage pathogens could cause if they acquired the appropriate mix of characteristics. But fear of this scenario should be tempered by the knowledge that this natural experiment cannot be repeated. Human pathogens are now so thoroughly and continuously mixed that we cannot generate the coevolutionary arms races in one area that would leave the humans who were not running in the race vulnerable when they eventually encountered the pathogens. No more New World populations of humans are left to discover, contact, and destroy.

Another problem with the designer bug strategy is the continued evolution of virulence and transmissibility after the bug is released. A key characteristic of the designer bug would be transmissibility, but once transmission occurs, the evolutionary future of the bug is out of the hands of the terrorists. It would rapidly evolve to a level of virulence that is most suited to the environments in which it is being transmitted. In other words, it would evolve into an organism with more familiar characteristics. The organism may cause terrible damage during the process of evolutionary equilibration and thereafter. But, as noted above, it would still leave the target population intact, ready to take revenge on the terrorists. The influenza pandemic of 1918 illustrates something like a worst-case scenario. Here was a pathogen that was probably overly exploiting its host during most of the pandemic; as it swept around the world, the most virulent viruses eventually lost out in competition with viruses of reduced virulence. The generation and maintenance of high virulence was permitted by natural selection only when the dependence of transmission on ambulatory hosts was temporarily relaxed under the unusual conditions of the Western Front and the masses of densely packed populations transported at the end of the war. It killed about 1 percent of humanity, yet it hardly seemed to influence the ability of the afflicted countries to keep the social cogs moving. Within a year or two the nasty variants of the influenza viruses vanished and never resurfaced in epidemic form. For more than eight decades, they have been replaced by the more traditional influenza viruses, which have been less lethal than the 1918 viruses by one or two orders of magnitude. The nasty strains did not disappear because of any heroic control efforts. Rather, they probably disappeared because they were too damaging for their own transmission under normal conditions. Terrorists would have little control over such a virus.

FROM WARFARE TO PREVENTION

The origins of human suffering have always been more infectious than has been supposed. By all indications this generalization still holds true. Like other advancements in science, this new understanding of disease has the potential both for damaging and for enhancing the quality of human life. The recognition that pathogens are prone to evolutionary change raises the question of what good or ill human society will make of this new knowledge.

First, it is important to remember that the newly recognized realms of infection are populated by chronic diseases and that the weaknesses of biological weapons are compounded as one moves from the acute to the chronic. Imagine how impotent a biological weapon would be if, like HTLV-1, it caused cancers in only one out of twenty-five infected individuals and did so sixty years after the targets were infected.

The shortcomings of biological weapons go a long way toward reducing the threat of their use, but they do not go all the way. The dangers must still be identified and guarded against to reduce the threat. If anyone could get away with conducting large-scale lethal experiments on humans to generate a vicious biological weapon, the new information would indeed be very dangerous. But with safeguards against such possibilities, the pros far outweigh the cons. To protect against the abuse of this new knowledge, we must make sure that such safeguards are in place, even if it means the abrogation of the sovereignty of any government that would dare to conduct experiments on humans that are designed to enhance virulence.

The world is not as safe as it could be, but it seems safe enough to make good use of the new biological knowledge. To take full advantage of the health benefits arising from the emerging understanding of the evolution of virulence, we need to adjust our environments to do the opposite of what a terrorist would do. We need to selectively disfavor transmission of harmful strains.

One of the greatest difficulties encountered by traditional approaches to controlling disease for good or ill is the flexibility of some pathogens. This flexibility is a difficulty for us if we are fighting against it—when, for example, we use antimalarials to try to suppress the versatile agents of malaria. But the flexibility of an infectious organism can also be a part of the solution to disease control.

CHAPTER THIRTEEN

The Protean Opponents

THE SEA NYMPH IDOTHEA TOLD THE LOST AND DESPERATE Menelaus what he needed to do to get back home to Sparta. Menelaus would have to force the instructions out of her father, the sea god Proteus. But Idothea warned Menelaus that the encounter would be challenging. Proteus would change his form to avoid being controlled. At one moment he might be a serpent, at another a leopard or a wild boar. If Menelaus could avoid being deterred by any of these forms, the sea god would eventually transform himself into a benign cooperative shape. By anticipating Proteus' changeability and developing a strategy for coping with it, Menelaus could control his adversary and resolve his predicament.

This allegory is illustrative of the way we must face the problem of infectious disease. Many once-devastating infectious diseases have been controlled so effectively that they are no longer a problem for most of the world's societies. Smallpox, diphtheria, polio, typhus, and measles are either no longer a problem or a problem only in areas in which the infrastructure and economic support are insufficient to implement proven solutions. But these diseases are the easy adversaries, the ones that could be controlled by identifying their Achilles' heel and attacking it with a primitive weapon. For some pathogens the Achilles' heel was vulnerability to the immunity generated from natural infection and from vaccines, a vulnerability that is attributable partly to a low potential for change. Smallpox, for example, had such a low potential for generating variation that a primitive vaccine made from a distantly related virus could rid it from the human population. We are left with the wily adversaries, protean opponents that change in response to our attacks.

HIV is one of these protean opponents. Its high mutation rate and potential for recombination allow it to wriggle free of the control we try to impose through our immune system and antiviral drugs. We have not even gotten close to controlling it with a vaccine.

Malaria is another protean opponent. Its flexibility arises from its sexual reproduction. The malaria organism's version of sperm and egg come together in the mosquito, reshuffling the genetic deck before each round of human infection. If several different genetic instructions are particularly useful to a malaria parasite, they can come together by way of this shuffling so that one organism can put together a hand combining the best that the trillions of parasites in a region have to offer. It might be a combination of instructions for countering one or more commonly used antimalarial drugs, or new versions of proteins that allow the organism to evade the immunity that has been acquired by the people in the region.

Ever since the antimalaria campaign fell apart in the 1960s, master planners have been dangling the prospects of an effective vaccine that could eradicate malaria. It is a desperate hope of last resort. Humans have never controlled a sexually reproducing protozoal parasite by vaccination. Vaccination against viral and bacterial diseases has a long record of success, but this success tells us little about the possibilities of controlling a sexually reproducing protozoan. Far from providing long-term control, antiprotozoal vaccines have not yet worked well even in the short run. For malaria control, working in the short run will be the easy part of the task because it will involve stimulation of the protection that already occurs among people who have acquired resistance from previous infections. If efforts at vaccine development continue, a vaccine that is moderately effective in the short run will probably be developed. But after its introduction, such a vaccine is bound to become steadily less effective as the variants that are not suppressed by the vaccine take over.

Why the blinders? Part of the reason is that immunological suppression of malaria occurs among people who live in malaria-infested areas. If these people are resistant, the thinking goes, then we should be able to generate similar resistance with the right vaccine. This reasoning focuses on the full half of the glass: a vaccine probably could be developed that would provide some degree of protection against malaria. The empty half of the glass, however, tends to be ignored: in spite of these moderately effective immunological responses and the genetic defenses that are present in people in endemic areas (such as alleles for sickle-cell anemia), the "resistant" people still develop malaria. As with HIV, there are already variants in the pathogen population that are not adequately controlled by the immunological response. Even without any novel variants arising, the existing variation suggests that vaccination would be at best moderately successful. As new variants arise and spread, the control will weaken further.

But desperation is another important cause of the reliance on the vaccine option. Researchers do not know where else to turn. They are convinced that the solution will come from some high-tech fix, and the only high-tech fixes they can think of involve the use of insecticides, antimalarial drugs, and vaccines. Insecticides cause evolution of resistance in insects; antimalarials cause evolution of resistance in malarial parasites. Vaccines will probably cause resistance to vaccines, but this barrier has not yet been encountered—so, the logic goes, let's try, and hope that it does not happen. In 1991, after several decades of research directed toward malaria vaccines, the London School of Hygiene and Tropical Medicine published an overview of approaches entitled *Malaria: Waiting for the Vaccine.* A decade later we are still waiting. Malariologist Kamini Mendis entitled the final chapter of the volume "Malaria Vaccine Research—A Game of Chess." A good metaphor. He did not mention, however, that human immune systems have never won such an immunological game of chess, whereas malaria has been honing its game against our immune systems for at least many thousands of years, and probably since before there were humans.

There is another option. Rather than trying to beat the master at his own game, change the game, and change the adversary. That is what we have done for all our domesticated animals and plants. Instead of being stymied by evolutionary responses to control strategies, incorporate the evolutionary responses into the strategies. For malaria, as well as for every other infectious scourge, understanding the reasons for the evolution of harmfulness and benignity points to interventions that will give people something that they want in the short run while making the pathogens evolve toward benignity in the long run.

Modern medicine uses a three-pronged method to control infection, each of which—hygiene, antibiotics, and vaccines—can be used with an eye toward evolutionary control. The evolutionary framework is like a handle that transforms the three prongs into a more effective tool, a pitchfork that can be used not just for attacking microbial enemies but also for giving them evolutionary pricks in the butt to make them evolve down a trajectory favorable to us.

That is how we have most successfully dealt with conflicts of interest between us and the animals and plants we depend on, and in fact we have already unknowingly applied the method to many pathogens to make them work for us. The polioviruses that have been attenuated in live vaccines, for example, are domesticated viruses. That kind of domestication is fairly easy to control because it occurs in dishes, tubes, and flasks. An experimenter can pick and choose organisms with desirable characteristics for the next generation. But the health sciences have been particularly unimaginative in restricting their efforts at domestication to the lab.

Selective breeding and genetic engineering of domestic animals and plants is now a major industry, but the first stages of domestication probably resulted from people doing things they wanted to do in the short term without any thought of long-term consequences; these shortsighted activities then changed the organisms people were interacting with over the long term. People who carried different wild seeds in pots for growing the following year would favor seeds that had characteristics that would allow them to keep well in pots. People who found a wolf pup cute and cuddly were more likely to take it home and became attached to it, and so on down the line, as that lineage of wolf evolved into a lineage of dog.

There is a lesson to be learned from the history of domestication. People acting in their short-term interest can be powerful forces shaping the organisms with which they interact. If we want to domesticate pathogens without taking them into the lab, we will probably be most successful if we nudge people down paths they would like to go in the short run. If we nudge ourselves in the right direction, the evolution of our pathogens will follow.

SCREENING FOR MOSQUITOES

The general solution for acute infectious diseases is to invest in interventions that reduce transmission from the sickest people. This investment should make pathogens evolve toward benignity because the sickest people tend to house the nastiest pathogens. The particular solution depends on the particular category of disease.

For diseases transmitted by vectors such as mosquitoes, mosquito-proof housing should push the host-parasite relationship toward benignity. When houses and hospitals do not restrict entry of mosquitoes, very sick individuals are particularly vulnerable to mosquito bites. When houses and hospitals are mosquito-proof, mosquitoes do not have access to a person sick in bed. The parasites in that person are therefore taken out of the competition. Transmission occurs instead from people who are sufficiently healthy to walk around outside. Because these people tend to be infected with more benign parasites than are severely ill people, mosquito-proof housing should tip the competitive balance in favor of the milder strains.

Mosquito-proof netting might seem like a reasonable and cheaper way to accomplish the same thing. But it is bound to be less effective because it demands more of the sick person and those who care for the sick. Can one count on a sick person to remember to tidy up the mosquito netting and to be motivated to do so? People generally use mosquito netting to avoid being infected rather than to avoid infecting others. Caregivers too tend to be focused on the patient rather than the long-term well-being of the society. They may be less inclined to be careful about mosquito netting if the person already has malaria.

Mosquito-proof housing obviates these problems because a sick person needs to do only what that person feels like doing: lying down and taking it easy. Similarly, the caregivers need focus only on their patients. Nor do the people involved have to put up with short-term inconveniences (shots or periodic spraying of houses) for a long-term gain. They simply have to be willing to accept something that they would like anyhow: higher quality housing.

Skeptics may argue that it cannot be done for a reasonable price. It already has. Just before the discovery of DDT and the synthetic quinine derivatives, the Tennessee Valley Authority was confronted with a malaria problem. The dams it had constructed during the 1920s along the border between Tennessee and Alabama had created mosquito-breeding areas; by the mid-1930s about half the Alabaman residents along the border had malaria infections. Faced with limited options, the TVA decided to embark on a major mosquito-proofing campaign. They divided the area into eleven zones and, between 1939 and 1944, mosquito-proofed every house at a cost of about one hundred dollars per house. Mosquito-proofing was maintained at a cost of about ten dollars per year per house (all 1940 dollars).

Skeptics may argue that screening of houses may do little to influence malaria transmission because much transmission may be occurring outside houses. The outcome of the TVA program provides some insight into this possibility. The mosquito-proofing was staggered so that some zones were completed sooner than others. By the time the mosquito-proofing was completed in a zone, the prevalence of infection had generally fallen to less than 10 percent. Within a couple of years after that, the prevalence was less than 1 percent in virtually every zone.

Skeptics may also argue that people will not tolerate the reduced airflow associated with screens. The TVA personnel were worried about this problem too in the stifling Alabama summers. They reported a few libertarians who threw furniture through the screens in protest, but overall they had virtually 100 percent compliance. In general the people were pleased to get the screens so that they could now escape the buzzing and biting of mosquitoes.

Skeptics may argue further that Alabama is not Nigeria. Differences between countries will undoubtedly influence the outcome of such interventions. In some countries the intervention will be more economically and politically feasible than in others, especially with regard to the initial intervention and long-term maintenance. The logical starting point would be to try the intervention in those countries that seem most appropriate and then to move on to assess the limits of the strategy.

Skeptics may still argue that we do not even know whether, after all the trouble, screening will have the desired evolutionary effect. Yes, we do need to try the intervention to see whether it will work. But this prospect is as close to a win-win situation as one could realistically ask for. If the evolutionary argument is wrong, and no evolution toward benignity occurs, then a large number of people will at least have obtained better housing and will have fewer mosquito bites and a lower rate of infection. If the malaria responds as the malaria in Alabama responded, it will be virtually eradicated. If the local malaria is not eradicated, the intervention will serve as a test of the evolutionary hypothesis.

If such tests prove the evolutionary argument correct, the death and suffering associated with malaria and other vector-borne diseases in the area may be dramatically reduced because each infection will be less damaging—one cannot say by how much without running the intervention, but the geographic variation in the harmfulness of malaria provides some indication. In areas with very restricted seasonal abundance, such as along the northern edge of sub-Saharan endemism for falciparum malaria, and in northern latitudes for vivax malaria, infections are relatively benign. Even where malaria transmission is intense, such as in Gabon, mild strains of the falciparum organism coexist with harmful strains, indicating two things. First, the raw material for evolution to benignity is there. Second, the mild strains, being already able to persist, might need only a slightly increased competitive advantage to dominate or even displace the more harmful strains.

Obviously I have been talking with some skeptics over the years. Perhaps the most frustrating response, though, is dismissal on the grounds that the intervention is primitive. This kind of dismissal is sometimes made by those who are convinced that the answer must lie in some technological breakthrough, even though mosquito-proofing of houses is supported by solid evidence, and the high-tech applications provide no evidence of an imminent breakthrough. The mosquito-proofing approach will undoubtedly incorporate high-tech advances in materials science and molecular methods in many aspects of the intervention, surveillance, and testing. But even if it did not, why care? What matters is what works, not whether it involves mesh, boards, and nails.

Finally, it is worth noting that the same intervention should simultaneously cause evolutionary reductions in other vector-borne diseases in the region. The experience with dengue along the Mexican border suggests that a mosquito-proofing campaign would reduce the damage from this viral disease as well. The cost-benefit ratio could be tremendously favorable.

WATER POWER

The evidence for evolutionary control of the harmfulness of diarrheal pathogens is at a more advanced state than the evidence for vector-borne pathogens. Evolutionary considerations suggest that provisioning of clean water supplies should cause an evolutionary reduction in the harmfulness of diarrheal pathogens. Indeed, much of the evidence that waterborne transmission causes evolutionary increases in harmfulness (discussed in chapter 1) comes from studies in which the provisioning of clean water supplies was associated with a replacement of harmful diarrheal pathogens with similar but milder pathogens. The most harmful agents of bacterial dysentery, for example, were replaced like clockwork by milder species in country after country as water supplies were purified. These studies indicate that the same trends would occur in poorer countries if sufficient investment was made in cleaning up water supplies. Does this change occur on a finer level within particular species of pathogens? If so, would the time period be sufficiently short to allow this evolutionary change to be incorporated into a control strategy? In theory such changes could be stronger and more rapid; the more similar bacteria are, the greater the intensity of competition between them, particularly because the cross-reactive immune response becomes stronger as pathogens become more similar.

One can imagine some large-scale experiments that could resolve the issue. For example, one could release a diarrheal pathogen like the agent of cholera, *Vibrio cholerae,* in one country with a poor water supply and in another country with a protected water supply and then follow the evolutionary trajectory in each over the next decade to see whether *V. cholerae* evolved reduced harmfulness in the region with the protected water supply. Obviously the ethical problems with such an experiment would make its rejection a no-brainer for a funding agency. The next best thing would be to see whether a natural version of this experiment had occurred and could be analyzed to find out if the predicted changes had taken place. In January 1991 such a natural experiment occurred. Cholera arrived in Peru and quickly spread throughout Central and South America, to countries with both good and poor water quality. Although this arrival caused and is still causing great difficulties for the inhabitants of Latin America, it created an opportunity to assess whether a disease organism like the agent of cholera could become rapidly more benign in response to transmission in an area with clean water supplies.

V. cholerae is particularly amenable to such study largely because its inherent harmfulness is quantifiable by measuring its toxigenicity, the amount of toxin it produces under carefully controlled growth conditions. The specific onset of the epidemic allows for the assessment of whether any evolutionary reduction in *V. cholerae*'s toxigenicity could occur in a time period comparable to that over which other kinds of interventions might provide beneficial effects. Specifically, if the toxigenicity of *V. cholerae* declined in areas with relatively safe water supplies but not in countries with unsafe water supplies during the 1990s, evolutionary control of virulence through improvement of water quality would occur over a time period comparable to the best of the alternative options, such as vaccination programs or interventions to reduce the frequency of infection. Since 1996 my colleagues in Latin America have sent strains *Vibrio cholerae* to our lab at Amherst to test this idea.

The key experiments were done in 1998 and 1999 largely through the hard work of two of my students, Alissa and Jill Saunders, a pair of identical twins whose effervescence and all-American charm made stepping into the lab during this period eerily like stepping into a commercial for Doublemint gum. They led a crew that tested about one hundred strains sent in from Chile, Peru, and Guatemala. The particular prediction was that the toxigenicity of strains isolated from Chile, where water supplies are relatively safe, should have declined during the 1990s; the toxigenicity of strains from Peru and Guatemala should not have declined as much, if at all, because water supplies in these countries have been less free of fecal contamination. The strains isolated from Chile in 1991 were variable in their toxigenicity; but strains with high toxigenicity were absent from collections made just a few years later. Not so in Peru and Guatemala, where a broad range of toxigenicity persisted throughout the decade. The corresponding difference in the incidence of cholera in Chile was particularly dramatic. In 1994 only one case of cholera was reported in Chile. In neighboring Peru twenty-five thousand were reported.

The entire set of findings on diarrheal diseases offers strong support for the idea that the diarrheal disease problem as a whole could be greatly ameliorated by an intervention that gives people what they would prefer to have anyhow. All else being equal, people prefer water that is not contaminated with fecal material. As is the case with mosquito-proof housing, people do not have to know about the long-term evolutionary benefits of the intervention, nor do they have to be forced to accept something they are averse to. They need only be offered ready access to uncontaminated water that is as aesthetically pleasing as the contaminated water.

The evidence from Latin America also lends support to the idea that the evolution of antibiotic resistance can be controlled by controlling the evolution of pathogen harmfulness. If increased harmfulness favors increased antibiotic usage, which in turn favors increased antibiotic resistance, then one should find the more harmful lineages within a region to be more resistant to antibiotics. The Guatemalan strains provide the best test of this idea because they vary widely in toxigenicity and because the antibiotic usage pattern is known. Trimethoprim-sulfamethoxazole was the drug of choice in Guatemala during the 1990s. In accordance with the theoretical predictions, the resistance to this drug was significantly correlated with harmfulness. By causing evolution toward benignity through the provisioning of safe water, one should be able to reverse the process, thereby favoring reduced antibiotic resistance as well. The data set as a whole is therefore encouraging with regard to the possibilities of controlling the evolution of both harmfulness and antibiotic resistance through water purification.

These ideas can be similarly applied to the other factors that allow transmission from immobile hosts. Mosquito-proofing of housing, for example, should favor not only reduced virulence of malaria but also reduced antimalarial drug resistance. Similarly, according to the logic presented in chapter 1, reductions in attendant-borne transmission in hospitals should favor the evolution of reduced harmfulness and reduced antibiotic resistance among the pathogens acquired in hospitals.

SAFER SEX

When I teach the introductory course in biology at Amherst College, I make a special effort to talk about the sexually transmitted pathogens that are circulating among sexually active college students. I go into considerable detail about the diseases these sexually transmitted pathogens cause and the relatively high frequency of infection among college students. This tour includes the standard fare of things that do not make it into many R-rated movies: AIDS, ugly lesions due to syphilis, yeast infections, and pus-laden or cheesy discharges. Of course, I cannot stop there. I go on to mention the less widely recognized damage from venereal diseases, such as T-cell leukemia, paralysis, infertility due to oviduct scarring, ectopic pregnancies, cervical cancer, liver cancer, Kaposi's sarcoma, and pelvic inflammatory disease. I also mention the high probability that there are many more chronic diseases that may be caused by sexually transmitted pathogens but for which there is presently insufficient information to label them sexually transmitted diseases. In this category are endometriosis, miscarriages, chromosomal damage of fetuses, low sperm counts, and penile cancer.

At the end of the lecture, when I ask how many will be less likely to have sex with someone they have been feeling ambivalent about, 90 percent or so of the students raise a hand. My obliging students may be humoring me, but the pallor of their faces suggests that they are answering the question honestly. Simply providing people with information about the way things really are alters mind-sets. The extent to which the altered mind-sets actually alter sexual behavior remains to be demonstrated by carefully controlled studies, but it would be very surprising if there were no effect. This little exercise is an indication of the kind of changes that could be enacted simply by providing information. For all the talk of education about the dangers from sexually transmitted diseases, almost no one grasps the full scope.

The short-term consequences of reducing the potential for sexual transmission are widely recognized, but the long-term evolutionary consequences are not. In chapter 2, I discussed the evidence indicating that sexually transmitted pathogens evolve to become more nasty when the potential for sexual transmission increases. The silver lining is the flip side of this association: if the potential for sexual transmission is reduced, sexually transmitted pathogens should evolve reduced harmfulness.

The harmfulness of sexually transmitted pathogens can be controlled by forcing their evolution toward reduced harmfulness within an area and by maintaining resistance to invasion of harmful pathogens from outside regions. Distinguishing the two mechanisms is useful in clarifying the two kinds of tests that can be run to evaluate the idea. The first kind investigates whether a given situation is resistant to invasion from the outside. The second keeps track of the evolutionary course of pathogens that entered at some time in the past. If the invading type is distinct, the first kind of test can provide an earlier indication of the success of the control measures. The second kind of test is more powerful but generally requires more time. Both kinds of tests have been inadvertently set up as natural experiments in different geographic regions, much like the experiment that was set up by cholera's invasion of Peru.

An example of the first kind of test involves HIV in West Africa. Harmful HIV-1 lineages have invaded areas where the less harmful HIV-2 had previously been present. If the mildness of HIV-2 really does reflect a low potential for sexual transmission, then one expects that the HIV-2 will maintain itself where the potential for sexual transmission is low, and HIV-1 will gain ground where the potential for sexual transmission is high. The latter trend has already been documented in countries such as the Ivory Coast and Guinea-Bissau, where HIV-1 has been displacing HIV-2. A critical test of the idea will arise from the experience in Senegal.

I first wrote about this general prediction in 1993. I cast it specifically in the context of Senegal early in 1997 in response to a leading HIV expert who predicted that the displacement of HIV-2 by HIV-1 that was occurring in some West African countries would soon occur in countries such as Senegal. I suggested that the situation be revisited in the year 2000. What happened? In contrast to its course in countries with higher potentials for sexual transmission, HIV-1 has *not* displaced HIV-2 in Senegal. This confirmation offers hope that a low potential for sexual transmission could help control the AIDS problem, not just by reducing the frequency of infection but also by favoring mild strains over harmful strains.

The test will continue. If the spread of HIV-1 has been checked in Senegal, then the kinds of health policies and social norms that have been responsible for a low potential for sexual transmission there may serve as a model for rich and poor countries alike. The relevance for countries that cannot afford the expensive antiviral treatment is obvious; the relevance for rich countries is apparent when one considers the side effects of expensive antiviral treatments and the disintegration of the treatments' effectiveness that will probably occur as HIV evolves resistance. Even now, though the antivirals look fairly effective by some statistics, such as years of life conferred, they are less so by statistics that take into account the numerous damaging side effects of the drugs.

The situation in wealthy countries also differs from that in West African countries because HIV-2 is generally absent in wealthy countries. Mild HIV-1 strains, however, should be even more effective than the mild HIV-2 strains at suppressing harmful HIV-1 strains because of their increased immunological cross-reactivity. The evolutionary theory therefore predicts the same result: lowered potential for sexual transmission should lower the harmfulness of HIV.

Tests of this prediction will tend to be of the second sort mentioned above. One such test is occurring in Japan. One of the nastiest of HIV-1 subtypes, subtype E, first entered Japan from Thailand around 1990. Because Japan has one of the lowest potentials for sexual transmission, evolutionary theory predicts that the subtype E lineages that have been cycling in Japan should evolve reduced harmfulness. This prediction can be assessed by comparing the markers of disease progression among those who are becoming infected now and who will become infected in the near future. The viruses that have had a longer evolutionary history in Japan can be distinguished from the recent entrants by their nucleic acid sequences. The long time between infection and disease hinders assessments of the harmfulness of the viruses, but indicators of disease progression can be obtained by making other measurements. Viral densities inside people during the first few months of infection, for example, are higher in people who will progress more quickly to AIDS. This indicator could be evaluated immediately and the eventual progression to AIDS followed. Development of test-tube assays for the inherent harmfulness of the viruses would make the job much simpler, though development of such assays seems to have been given low priority among funding agencies.

The geographic associations between the potential for sexual transmission and virulence cut across the entire spectrum of sexually transmitted pathogens. Evolutionary interventions that reduce this potential for transmission may therefore be especially cost-effective solutions because the interventions that favor reduced harmfulness in one pathogen should as a rule simultaneously favor reduced harmfulness of sexually transmitted pathogens.

STAYING HOME

If sufficient information about the evolution of harmfulness is known and acted on, this approach could eventually lead to pathogens so mild we hardly notice the disease they cause. Imagine the effect of a simple change in our health ethic. What if people were encouraged to stay home if they or a family member were sick?

In today's workplace, someone who does not come in because of illness may be seen as being less dedicated, even though a sick worker may spread illness to the other employees. The advertising of products for symptomatic relief must exacerbate the problem, encouraging us, as they do, to spread our nasty respiratory tract infections. Cold and sore throat remedies are sold with promises to help you make it through your daily routine in spite of having a bad cold or sore throat. We are shown some uninformed victim in an unattractive state, all stuffed up and achy. Then we see the same person out and about after a dose of the peddled product, in the workplace, shopping, or taking care of neighborhood kids. That person, of course, is still infectious and is comfortably spreading misery throughout the community. The advertisements do not actually say, "Feel better fast so that you can spread disease within your community," but that is certainly their short-range effect. Their long-range effect is just as disturbing. The advertisements encourage the spread of nasty variants, whereas from an evolutionary point of view we need to do just the opposite. If people stayed home when they felt the least bit sick, the pathogens that would be left to circulate in the community would be those that rarely, if ever, made a person feel the least bit sick.

The necessary social change should not be too hard. It is in our nature to take it easy when we feel sick, and it is probably even beneficial financially to employers, though not to the companies pushing the get-out-of-bed-and-spread option. We have done this sort of thing before with smoking and drinking. We have countered the irresponsible advertising with responsible messages. Within a decade or two, public attitudes toward drunk driving and smoking changed from permissiveness to social ostracism. Countless lives have been saved. Changing attitudes toward coming to work sick may be easier because we have an inherent revulsion to infectious disease. You don't have to be taught to be disgusted when someone sneezes or coughs in your face. It should be relatively easy to get across the idea that a sneeze from a person in the chair next to us is almost as dangerous as a sneeze in the face; either may force us into days of bedridden misery. Sometimes it can even mean death, as is the case for elderly people infected with influenza. I doubt that most people would object too strenuously to staying home if they were not going to be criticized for doing it, especially if they would face the opposite: ostracism for going to work sick.

From the government and health experts there is silence, perhaps because the scope of the problem is not recognized. This view of the future might have been a pipe dream ten years ago, but today people can often do their work at home and send it in over the Internet. As is the case with drunk driving and smoking, even partial compliance should provide a benefit. The more compliance, the greater the benefit. As is the case with waterborne, vector-borne, and attendant-borne transmission, solving the harmfulness problem should simultaneously ameliorate the problems of antibiotic resistance. Recruit your bosses. If they are unresponsive, punctuate your argument with a well-placed sneeze.

CHAPTER FOURTEEN

Tools of Domestication

ALTHOUGH THE CONQUEST METAPHOR HAS LIMITED APPLICAbility in the control of infectious diseases, the diversity of warfare situations offers a diversity of metaphors. The matches should not be surprising, because in a very real sense we are in a state of war with many of our microbes. They *are* invading us. We *are* killing them. The mistake in the standard war metaphor was in thinking that there was only one successful solution: unconditional surrender. In real warfare as in antimicrobial warfare we have learned, sometimes very painfully, that there are other solutions.

Consider the Soviet war in Afghanistan. Although the Soviet Union had overwhelming superiority in fire power, the United States saw that it could tip the balance in favor of the Afghan rebels by providing Stinger missiles, which could knock out the main Soviet threat: attack helicopters. The United States provided this weapon after considering which of the two sides was more dangerous to U.S. interests. There was no opportunity to groom a successor. If there had been, the U.S. government might have called on the CIA to deploy a different strategy. Whether one supports or condemns any such covert action depends on how one weighs the expected geopolitical benefits against the compromises of individual rights and national sovereignty. The quality of the assessment depends on the range of strategies considered.

Health policy strategists have a wider range of opportunities because bacteria and viruses do not have rights. They are free to use any strategy regardless of the destruction that is imposed on innocent microbes. Yet in spite of this freedom of action, health strategists have not drawn on the range of tactics the CIA employs. Imagine how much less effective it would have been if the United States had intervened in Afghanistan by battering down both sides. Yet that is just what vaccination programs attempt to do. Instead of making vaccines that favor the mild strains by selectively knocking out the truly dangerous opponents, or by grooming the microbial successor, we have been trying to make vaccines for over two centuries that knock out all the variants of a target microbe, whether the variants are mild or extremely dangerous. When a vaccine does that, we no longer have the mild variants predominating in the wake of the campaign to protect against the harmful variants.

If vaccine efforts lapse, as they often do, then the mix of pathogens that was there before the vaccine effort will quickly expand to fill the void. If only mild variants are left, the situation is more stable against reinvasion by harmful variants. The overall kill rate is lower, but it is selective; the outcome is thus more favorable.

The second most cost-effective vaccination program in history, the one that controlled diphtheria, inadvertently showed how well this selective strategy can work. The people who made the diphtheria vaccine may have thought of their efforts as an all-out war to eradicate an enemy, but the bacterium that causes diphtheria, *Corynebacterium diphtheriae,* was not eradicated by the vaccine program. Rather, the vaccine selectively suppressed the dangerous competitor, altering the balance in favor of the benign competitor. This selective intervention virtually banished diphtheria for more than a half century without the need for an all-out eradication campaign. If we understand why this campaign worked so well, we might use it as a model for other vaccine campaigns.

The diphtheria bacterium causes most of its damage as a result of a toxin it produces when it is short on resources, particularly iron. The toxin costs the bacterium about 5 percent of its protein budget, but the investment pays back dividends because the toxin kills the cells of the respiratory tract near the bacterium, thereby liberating the nutrients the bacterium needs. The diphtheria vaccine was made by modifying this toxin a little so that it no longer damaged respiratory tract cells but still caused the immune system to generate antibodies that would recognize and sequester the unmodified toxin. If a toxin-producing *C. diphtheriae* invades a person who has been vaccinated, the toxin is sequestered by antibodies before it can destroy a person's cells and provide nutrients for the bacterium. The 5 percent cost of toxin is simply a drain on the bacterium's ability to compete with toxinless bacteria. The overall effect is that the strains that do not produce the toxin win out over the harmful strains. Wherever the strains left in the wake of a diphtheria vaccination program were assessed, the same trend occurred: the toxin-producing strains vanished, replaced by the milder, toxinless strains. That is a good outcome for us because strains that do not produce toxin not only fail to cause diphtheria but also protect us against the harmful strains that do. They therefore act like free live vaccines.

These arguments lead to a simple rule for vaccine development. Whenever possible, use virulence antigens: those components of a pathogen that make viable, benign organisms harmful. Doing so will generate an immune response that selectively protects against the harmful organisms. Including antigens against components of the pathogen that do not make it virulent must be avoided. Otherwise the vaccine will remove mild strains that could further suppress the harmful strains.

This virulence antigen strategy has been used inadvertently in one other vaccine program, the one against *Hemophilus influenzae,* which has been an important cause of encephalitis in children. That program was so successful that it left researchers scratching their heads. But extraordinary success is what one should expect from virulence antigen vaccines. The strategy should be applicable to all vaccines, yet it has not been considered as part of the strategy for making *any* vaccine, largely because vaccine developers tend not to look at their task from an evolutionary point of view.

The virulence antigen strategy requires that vaccine experts shift away from eradication as a goal. This shift is dictated for some diseases by the ability of vaccines to prohibit disease but not infections. When children are vaccinated against pertussis (whooping cough), for example, the disease is generally prevented but the organism is still present and transmissible. Prospects for eradication by such vaccines are obviously very dim, no matter how pervasive the vaccination program. We can expect to be living with the agents. If we have to live with the organism anyhow, we should make it a benign organism that supplements vaccination efforts rather than a mix of largely harmful organisms. Pertussis vaccination is a perfect candidate for a virulence antigen strategy, not just for this reason but also because virulence antigens are already identified and can generate a protective effect that is comparable to the best vaccines available. The pertussis vaccines that are currently being used have other antigens, particularly one called filamentous hemagglutinin, which trigger immune responses that suppress benign strains as strongly as harmful strains.

VENEREAL VACCINES

The two-hundred-year history of vaccination has generated only one success story against a sexually transmitted pathogen: the hepatitis B vaccine. This program was put in action during the early 1990s. It looked good at the outset but is already showing signs of trouble. Hepatitis B, like HIV, can generate variation quickly through mutation. Recent data indicate that a massive hepatitis B vaccination program in Taiwan is causing the virus to evolve around the vaccine.

Why have vaccines not been as successfully applied against sexually transmitted pathogens as against other kinds of pathogens? Each sexually transmitted pathogen has its own story, but there are some general reasons that the standard approach to vaccination is not particularly effective against venereal diseases.

Sexually transmitted pathogens are the beneficiaries of several psychological idiosyncrasies. First there is morality. People have little control over whether they will inhale air laden with viruses, whether a mosquito will get lucky, or whether some of their food has been contaminated because of neglected hand washing. But people are supposed to have some control over whether they engage in sex. Treating syphilis with an antibiotic is remedying a past moral lapse, but vaccinating people against syphilis can be seen as encouraging immoral behavior. Allocating money for the development of venereal vaccines can be politically risky for the same reason.

There is also the immunological problem. Sexually transmitted pathogens have to be able to avoid destruction by the immune system, and vaccines rely on the immune system to destroy sexually transmitted pathogens. This is the problem that seems to be making the hepatitis B vaccine program show hairline cracks. Hepatitis B stays ahead of the immune system by mutating its form. If the vaccine primes the immune system to combat one form of the virus, the door is left open to another form.

Vaccines against cancer have generated much hope but little success in the way of disease control. If most cancers turn out to be caused by viruses, all of this might change. Consider cervical cancer. Once cervical cancer became recognized as an infectious disease, new opportunities arose. The papillomavirus that causes this cancer offers several distinct antigens that would not have been available as immunological targets if the cancer had been caused solely by mutations. Even with virally caused cancer there is the problem of getting the immune system to knock out a lump of the body's own cells, but at least there is a chance that the cells are hanging the virus's body parts as a KILL ME sign for the immune system. Cells infected with the most deadly forms of papillomavirus do seem to post such signs using parts of the E6 and E7 proteins; and cells that were engineered to express these proteins were destroyed by the cytotoxic T cells that would need to be mustered to take care of these cells in the body. There is still the problem of killing a big lump of cells from the outside. Once the lump gets too big, no amount of immunological activity may be sufficient. But the immunological action triggered by an E6 or E7 vaccine may work much earlier in the course of infection, long before a distinct lump is seen.

If such a vaccine does act against infected cells, it could have a much greater long-term punch than its developers suspect. The E6 and E7 proteins are the ones that sabotage the cell's ability to control its division. A vaccine derived from them would therefore be a virulence antigen vaccine that might provide an extra evolutionary punch by tipping the competitive balance in favor of the papillomaviruses without the damaging form of these proteins. These mild strains of papillomavirus would then be left to generate additional protection against the harmful strains by triggering immunity to antigens that the mild and harmful strains share.

AIDS vaccines present a different set of problems and opportunities. On the positive side, the infected cells are not cloistered inside tumors. On the negative side, the virus is so mutation prone that a vaccine that would protect against all variants is probably unattainable. Still, the virulence antigen strategy may provide substantial protection if it is used early in the infection as a therapeutic vaccine. This kind of usage requires that the most damaging forms of HIV's proteins be identified and used in the vaccine. When these forms then arise by mutation, they may be quickly knocked out before they have a chance to gain a foothold. The protein HIV uses to attach to and enter cells, for example, typically changes during the course of infection, allowing HIV increasingly to enter and damage the helper T cells. This change is critical because these T cells tend to be far more important to the orchestration of the immune response than are the other cells HIV infects when it does not have this altered form. If the altered form of this protein was used in a vaccine early during infection, the more damaging viruses might be controlled longer because the immune system would be ready for them when they arose. In this case the normal immunity generated by the less damaging HIV in the body is supplemented by a vaccine-induced immunity that specifically suppresses the more damaging form. The more of the harmful forms that could be included in such a therapeutic virulence antigen vaccine, the longer the delay in the breakdown of the immune system.

THE SISYPHEAN CONTROL OF ANTIBIOTIC RESISTANCE

The main emphasis of control of antibiotic resistance has been on reducing the use of antibiotics. A complementary option is to restrict use according to the transmissibility of infection. This strategy requires a detailed understanding of each pathogen's life history. Consider zoonotic pathogens—that is, pathogens that use other species as their natural "reservoir hosts." When those pathogens are not transmissible from humans, options arise for preserving the usefulness of antibiotics indefinitely. Lyme disease, toxoplasmosis, pneumocystis pneumonia, Rocky Mountain spotted fever, and rabies are just a few of the diseases that fall into this category, which is characterized by dead-end human infections. The category may be much bigger than currently recognized, including, for example, such unlikely diseases as breast cancer.

If we do not have a special fondness for the reservoir host—a rat, for example—then the reservoir host can be exterminated. If we are fond of the host—a horse or a cat, for example—then we can vaccinate the animal or treat it with antibiotics. But if we choose vaccination or antibiotics, there is a subtle but important guideline we need to follow. The antibiotic or vaccine that is used for the reservoir host should not be the same as the one used for the human.

The reason for this guideline is most apparent with regard to antibiotics. If in a dead-end zoonotic disease we restrict the use of an antibiotic to human infections, we have a good chance of burying the problem of antibiotic resistance. Antibiotic resistance may begin to evolve in the treated person, but if the infection is a dead-end infection, those resistant organisms will not get out of the person to evolve higher levels of resistance. If we restrict those antibiotics that are most effective in humans to humans, then the only antibiotic resistance we will need to deal with is the antibiotic resistance that can evolve during the course of a single infection.

This guideline requires that medical research be more comprehensive in its development of new antibiotics. We cannot have just one effective antibiotic for a given pathogen. We need at least two and preferably several. The antibiotic that is best for treatment of human infections needs to be reserved for human infections. The others can be made available for treatment of the reservoirs. We should have more than one of these others because the antibiotic will tend to evolve resistance in the reservoir host by the now familiar scenario. We will therefore need to switch to new antibiotics as each becomes less effective. The important benefit of this strategy is that we would always have some antibiotic that provided protection for humans.

This strategy can be considered Sisyphean control, in reference to Sisyphus, the mythic Greek who was condemned by Zeus to eternally roll a rock uphill only to have it roll down again before reaching the hilltop. By this strategy we condemn pathogens in humans to be continually beginning but never culminating their evolution of antibiotic resistance.

The strategy represents a shift away from broad-spectrum antibiotics for humans to narrow-spectrum antibiotics, which can be restricted more easily both logistically and ethically. This shift might have posed a problem or at least an inconvenience in previous decades because a physician may prefer not to wait for a particular diagnosis but to go ahead with a broad-spectrum antibiotic able to knock out a variety of probable offenders. But things have changed. The disadvantages of broad-spectrum antibiotics are becoming apparent. Paramount among these disadvantages is the short life of the antibiotic due to the more rapid evolution of antibiotic resistance that occurs when an antibiotic is used broadly. Perhaps more important, however, we are nearing a period in which the identification of a specific pathogen will be made in minutes rather than in days because tests will rely on detecting small amounts of pathogen and therefore will not require the time-consuming process of culturing the agent. With rapid diagnosis, the appropriate narrow-spectrum antibiotic can be identified quickly and used.

The argument for Sisyphean control by vaccines is essentially the same. If we use the same vaccine for humans and the zoonotic reservoirs, we will be putting selective pressure on the pathogens to evolve around the protection we are trying to provide in humans. This point about vaccines is not widely appreciated, but the evolution of vaccine resistance is almost as inevitable as the evolution of antibiotic resistance. Experience informs us, however, that vaccine resistance generally takes longer to develop, presumably because the immune system may rely on several different weapons that it generates in response to vaccines, and can change the weapon to some extent if the pathogen changes. Still, the microbial record of evolutionary change that has been recorded in the medical literature shows that vaccine resistance has evolved, for example, among strains of influenza virus, hepatitis B virus, and the bacterium *C. diphtheriae.*

Sisyphean use of vaccines and antimicrobials can be especially powerful because of different thresholds that must be met for humans as opposed to animals. Although some pet lovers would disagree, the general sense is that the safety standard would not need to be as high for animals as for humans. If a human vaccine has adverse events in one in ten thousand vaccinees, it is a vaccine that will need to be improved. The same frequency of side effects may be considered acceptable in animals, and not just because we value the lives of our domestic animals less than we value human life. Some kinds of damage may not be as important to animals. Imagine, for example, that the only damage a vaccine caused in humans was a very slight amount of brain damage—enough, say, to lower an IQ score by five points in one out of every hundred children. This amount of damage would probably be considered intolerable for a human vaccine. But would we be so concerned if one out of every hundred Felixes took an extra second to figure out how to sneak up on the kitchen table? There will be some who will adamantly proclaim, "Yes," but I think that most would consider it an acceptable cost to pay for better long-term protection of people against Felix's pathogens.

The same logic applies to antibiotics, though the threshold for safety is not so high. To be considered acceptable, the average negative effect of a vaccine must be below the average negative effect of the disease against which the vaccine protects, averaged over all vaccinated persons. In contrast, the negative effect of an antibiotic must be less than the beneficial effect of the antibiotic for the treated patient. The latter threshold is much less stringent.

The double standard may seem harsh to animal rights activists, but the stakes for human health are high. We are not just talking about common colds. The spectrum of protection might involve life-changing illnesses, such as schizophrenia, bipolar disorder, chronic fatigue syndrome, toxoplasmosis, and breast cancer.

Though the Sisyphean strategy is easiest to visualize for zoonotic diseases, it may be increasingly applicable in subtle ways to diseases that are regularly transmitted from person to person, particularly chronic infectious diseases because they may often have a nontransmissible end phase. The point is well illustrated by atherosclerosis and one of its probable causal agents, *C. pneumoniae.*

Transmission of *C. pneumoniae* between people may be occurring primarily or even entirely from acute lung infections, and those bacteria found systemically in atherosclerotic plaques may be causing dead-end infections. If so, the restriction of antibiotics to these infections could control them indefinitely without generation of antibiotic resistance. For this possibility to be realized, research needs to resolve an uncertainty and ratchet up antibiotic development, and do both things before *C. pneumoniae* is shown to be a cause of atherosclerosis.

The uncertainty concerns the degree to which the *C. pneumoniae* that are causing systemic infections are transmissible to other people. The distribution of immunological positivity to *C. pneumoniae* suggests that systemic infections may be relatively or wholly nontransmissible. Few children show immunological signs of infection until they begin attending school, at which time the prevalence of infection rises rapidly with age. About one third of the parents of these children have some atherosclerotic damage. If *C. pneumoniae* causes atherosclerosis and if the systemic infections of these parents were readily transmissible, one would expect to see more children who are not yet school age infected as a result of transmission from their atherosclerotic parents.

If systemic infections are not transmissible, then the effectiveness of antibiotics against them may be preserved indefinitely by keeping the antibiotics from being used for treatment of the transmissible pulmonary infections. If that is done, we can expect some resistance to evolve within some of the treated individuals. But if systemic infections are not transmissible, these antibiotic-resistant variants would not be transmitted and the antibiotic resistance would not have a chance to increase progressively from infection to infection.

If, however, such antibiotics are used indiscriminately for treatment of both pulmonary and systemic infections, we can expect the now familiar stepping-stone development of antibiotic resistance: the resistance that evolves during the treatment of respiratory tract infections can increase progressively, thus negating the efficacy of treatment of both pulmonary and systemic infections. The result would be that the most important antibiotics—those that provide effective treatment of the damaging systemic infections—would lose their effectiveness through the evolution of antibiotic resistance.

Given the existing uncertainties about the role of *C. pneumoniae* in atherosclerosis, it may seem premature to consider the control of cardiovascular damage by controlling the resistance of *C. pneumoniae* to antibiotics. If, however, this topic is postponed until *C. pneumoniae* is generally accepted as a primary cause of cardiovascular disease, it may be too late. As soon as a causative role for *C. pneumoniae* is demonstrated, we can expect a major increase in antibiotic treatment of pulmonary infections in order to reduce the chances that systemic disease will develop. Any antibiotics that are effective against systemic infection would need to be known by this time so that their efficacy could be preserved by restricting their use to treatment of systemic infection. Physicians cannot be expected to withhold a particular antibiotic for pulmonary infection based on the speculation that it might be effective in systemic treatment. If it is *known* that the antibiotic is effective in systemic infections and that systemic infections are not transmissible, the argument for withholding the antibiotic would be compelling, as long as there were other antibiotics available for pulmonary treatment. Knowledge about the transmissibility of *C. pneumoniae* would therefore be most useful if it is acquired before the role of *C. pneumoniae* in atherosclerosis is resolved and before the discovery of antibiotics that are effective against systemic infections.

This point raises the other urgent need: the development of an array of antibiotics that includes at least one that is just as effective against pulmonary infection as is the antibiotic that is to be reserved for systemic infection. The antibiotics used for systemic and pulmonary treatment should have different mechanisms of action so that resistance to the antibiotic used for pulmonary infection would not generate cross-resistance to the antibiotic for systemic infection. A similar argument applies to the other leading bacterial candidate for infectious causation of atherosclerosis: *Porphyromonas.*

This analysis may seem like overkill. Why so much research effort on a speculative argument? The reason is that the chronic damage that may be caused by these pathogens—atherosclerosis, stroke, Alzheimer's disease, and multiple sclerosis—eclipses the damage caused by any other known or suspected infectious process. Even when the Sisyphean strategy may not apply with full force, it may make the resistance problem more manageable. This situation may occur when pathogens are extremely mutation prone and cause long-lived transmissible infections that can be suppressed but not stopped by an antibiotic. HIV falls in this category. There now exists a battery of more than fifteen antiviral compounds that are effective against HIV. Almost all of them generate resistance in HIV when used alone or in paired combinations. The beginnings of resistance to the combination therapy of three or more drugs is now increasing.

The resistance that develops to some of these antivirals does not handicap the virus as much as the resistance that develops in response to others. Viruses resistant to zidovudine (AZT), for example, often appear just as viable in the absence of zidovudine as the nonresistant viruses. Viruses resistant to lamivudine, however, seem unable to resist the immune system as effectively as sensitive viruses. A drug like zidovudine is best saved for the late stages of infection, when the chances of transmission are low. Any resistance that develops during this period is therefore unlikely to be transmitted. The longevity of zidovudine in the society at large would therefore be prolonged. HIV's progress toward increased resistance to zidovudine in this case is cast as a Sisyphean task.

The success of multiple antiviral therapy seems to have lulled policy makers into a false sense of security. Now that this kind of therapy seems to be working fairly well, efforts to suppress resistance are treated less urgently. But the problem of resistance needs to be addressed before the development of multiple antiviral resistance. If it is addressed afterward, another opportunity to control resistance will have been lost.

THE EVOLUTIONARY COURSE OF THE CHRONIC

Guiding the evolutionary course of chronic diseases may seem more difficult than altering the course of acute infectious diseases. If the damage from the chronic disease has not evolved as an adaptation, then one might wonder how one can alter the organism by invoking adaptations. The trick is knowing where to look for opportunities. If a pathogen cannot be transmitted from a chronic infection, then this lack of transmission can allow for escape from counteradaptation, as discussed regarding antibiotic resistance among infectious agents of atherosclerosis. If transmission can occur, the trick may be to learn how to use the immune system as a device to control evolution that occurs as a result of competition within hosts, as with the use of therapeutic vaccines and antivirals against HIV. But whatever the particular details of the situation, one thing is certain: the pathogens will evolve, and we will cut off our options for controlling them if we neglect their evolutionary potential.

CHAPTER FIFTEEN

The Prepared Mind

IN 1972 THE *NEW ENGLAND JOURNAL OF MEDICINE* PUBLISHED a little essay packed with insight. Its author, Lewis Thomas, divided medical care into three categories. His first category, "nontechnology," comprises supportive care, which "tides the patient through diseases that are not, by and large, understood." In this category he placed the care given to patients with "intractable cancer, severe rheumatoid arthritis, multiple sclerosis, and advanced cirrhosis." He also included "a large amount of what is called mental disease." Thomas called his second category "halfway technology"; it comprises the patch-job solutions to the problems at hand. His third category was "decisive technology . . . the kind of technology that is so effective that it seems to attract the least public notice; it has come to be taken for granted." The iron lung for polio patients was halfway technology; the polio vaccine was decisive technology. The surgical removal of infected lung tissue of tuberculosis patients was halfway technology; the cure of tuberculosis with isoniazid and rifampin was decisive technology.

Thomas's words have been weathered by a quarter century of halfway technology, but his insight is still timely. Organ transplants, angioplasty, and drug treatment for mental illnesses are today's halfway technology. The decisive technology that will replace them has not yet been invented. Thomas's insight into cardiovascular disease is eerily relevant today: "When enough has been learned to know what really goes wrong in heart disease, one ought to be in a position to figure out ways to prevent or reverse the process, and when this happens the current elaborate technology will probably be set to one side." We're not there yet but, if preventable infections prove to be the primary cause of heart disease, we could be there within a decade or two.

With the clarity of hindsight, we can see that Thomas, writing in the early 1970s, may have been overly influenced by the association between his three categories of medical care and the sophistication of the technology that was then used to provide the care. We now know that a great deal of sophisticated technology can be mustered to provide care that would fit in his nontechnology category—sophisticated technology is used to develop drugs that provide only palliative care and to diagnose and monitor illnesses for which there is only palliative care. Patch-job solutions may involve technological advances that are far more sophisticated than the technology that solved major medical problems—compare the sophistication of twenty-first-century open-heart surgery with the sophistication of the eighteenth-century technology that allowed the creation of the smallpox vaccine.

One of Thomas's messages is that we should never be content to see the patch-job technology as anything more than a stopgap measure, a finger that must be put in the dike until the decisive technology is found. He emphasized that decisive technologies are not only better; they are also cheaper: "I cannot think of any important human disease for which medicine possesses the outright capacity to prevent or cure where the cost of the technology is itself a major problem. The price is never as high as the cost of managing the same diseases during the earlier stages of nontechnology or halfway technology." Another quarter century of experience has only strengthened his argument.

The problem is that decisive technology is hard to envision before it has arrived. The science fiction of the 1950s provides a sense of how easy it is to miss the major advances of the future, even in a genre preoccupied with envisioning future possibilities. One of the most prescient science fiction writers of this period was Robert Heinlein, who often used his stories as a vehicle for communicating insights about the technology and lifestyles of the future. In his 1951 collection of short stories, called *The Green Hills of Earth,* Heinlein wrote about space shuttles, space stations, lunar landers, hydroponics, and the recycling of nutrients in closed systems. Like other science fiction writers of the time, he had computers playing a major role in the technology of the future. And, like the other science fiction writers of the time, he missed the major technological innovation that would unleash the electro-information society of the late twentieth century: microminiaturization. In one of his stories a space pilot has an electronic calculator, but not one sufficiently powerful to calculate a new flight orbit, which has to be done by "tons of IBM computers" on a behemoth space station that orbits above New York. Another story in this collection describes a premier space agency, which uses the latest in electronic technology, including cellular phones and video phones, to help plan for a meeting of representatives from the other planets in the solar system. Not bad for someone writing around 1950. But the agency also relies on punch cards and slide rules!

In *Canticle for Leibowitz,* published in 1959, Walter Miller had a character trying to fix a computer at a time when the planets of nearby solar systems had been colonized. Miller insightfully had the computer being used for word processing and even had it making word-processing errors that bring back memories to those of us who are over 40. Miller missed the idea that computers with far more power than those in his room-filling cabinets would fit in a shirt pocket long before humans first set foot on their second planet.

The futurists could envision certain kinds of miniaturization. In his 1953 novel, *The Second Foundation,* set thousands of years in the future, Isaac Asimov had people using nuclear engines the size of a peanut. But he also had them using the Analytic Rule, a souped up version of a 1950s slide rule. It was the extreme miniaturization of electronic circuits that caught the futurists by surprise.

It was not until the 1970s that science fiction writers finally caught on to the idea that computers would become more powerful and much smaller. In their 1974 novel, *The Mote in God's Eye,* Larry Niven and Jerry Pournelle gave their society a pocket computer, which was used for calculation and dictation and could be linked to other computers to share information. Not surprisingly, Pournelle's other avocation was computer jock. If anyone could have foreseen the enormity of electronic miniaturization, it should have been the computer jocks, but even they seem to have seen it only when it was imminent.

It would be informative to make similar assessments of the technological advances that were envisioned in medicine by authors of the 1950s. But there is not much to assess, and what there is would not be fair to assess. Science fiction writers, who had the imagination, did not tend to have the medical knowledge necessary to articulate the future possibilities of medicine; and medical people, who had the knowledge, did not tend to write science fiction (at least not intentionally). The term *medical jocks* is not in the vernacular.

The future of disease control envisioned by medicine at mid-century lacked imagination. It tended to be more an embellishment of what was already there by 1950: new vaccines and drugs for the elimination of disease, and new gizmos for diagnosing disease. So what are we left with at the beginning of the twenty-first century? More of the same kind of thing that we had in the middle of the twentieth century. Some excuse this lack of innovation by arguing that all of the truly innovative approaches—antibiotics, vaccines, and hygienic improvements—had been discovered by 1950.

This conclusion reminds me of the frustration my father shared with his younger brother when they were growing up in the 1920s on a subsistence farm near Oshkosh, Wisconsin. He was discouraged because they wanted to be inventors, but all of the great inventions had already been invented. He was no dummy. (He eventually obtained a Ph.D. in physics and became a member of the faculty at Northwestern University.) Rather he just was not, as an adolescent, able to imagine the possibilities that existed beyond the mechanical and electrical innovations of his time.

Lewis Thomas dealt with this barrier by emphasizing that we have to look for decisive innovations in an open-ended way. He ended his essay by directing us to a general source of answers. "I would regard it as an act of high prudence to give high priority to a lot more basic research in biological science. This is the only way to get the full mileage biology owes to the science of medicine." I applaud Thomas for getting the problem right, but he addressed only part of the answer. Funding basic biological research is not enough to ensure that medical advancements will be timely. There is another critically important ingredient for the biomedical brew. Louis Pasteur referred to this ingredient in his statement, "Fortune favors the prepared mind."

Pasteur's dictum is as true today as it was in the nineteenth century, but its full relevance to medicine has been only partially appreciated. It is mentioned in the context of particular observations, such as Alexander Fleming's discovery of penicillin; his prepared mind led him to notice and wonder about the inability of bacteria to grow close to the *Penicillium* mold in his culture dishes. Fleming's recognition of the importance of a prepared mind is evident in his response to a critic who suggested that the discovery was just luck: ". . . the spores didn't just stand up on the agar and say, 'I produce an antibiotic you know.' "

Besides recognizing the significance of individual observations, however, prepared minds must have a conceptual breadth that extends beyond the prevailing dogma. Barry Marshall had a prepared mind when considering the causation of ulcers. This kind of prepared mind was a limited resource among ulcer researchers during the last half of the twentieth century, so much so that medical texts failed even to mention the possibility of infectious causation of ulcers from the late

1940s, when antibiotics were used effectively in New York hospitals, until the 1990s, when the Australians' rediscovery of infectious causation was finally generally accepted. One wonders how much longer it would have taken to discover the infectious causation of peptic ulcers had these doctors not come on the scene. If it had taken another forty years, we would see ulcers being cured with antibiotics only in the year 2030.

Some might be tempted to interpret this history in a way that is not so damning of the status quo. This line of argument might admit that the medicine of the 1950s may have been myopic but not the medicine of the 1980s and 1990s. After all, the infectious causation of ulcers was accepted in the 1990s. That in itself could be construed as evidence of a more insightful medicine in recent decades. The published record provides documentation that allows us to assess whether this view is accurate or whether it is a self-serving rewriting of history to reflect how it should have been rather than how it was.

If acceptance of the idea had proceeded along this favorable model for biomedical progress, we would expect such an important breakthrough to have been published in a leading journal. That publication should then have generated a flurry of independent tests to confirm or refute the purported finding. Upon confirmation, the breakthrough would have been accepted by the medical establishment. That is not quite how it happened. Marshall and his colleagues published the first part of their seminal work in 1983, and almost nobody paid any attention. Then, in 1987, they published one of the most critical tests of the idea, and again almost nobody paid any attention. What finally made people pay attention? What led to the scientific assessments that eventually established infectious causation of ulcers in mainstream medicine during the mid-1990s? Marshall remembers that interest began to be stimulated among researchers not as a result of consideration in the medical literature, the route by which these kinds of breakthroughs are supposed to advance medicine. Rather, Marshall and his colleagues credit a story in the popular press for getting the idea noticed and moved along toward acceptance. The story was published in the *National Enquirer.* Page 3 of the March 13, 1990, issue contained two stories. The top three quarters of the page was headed DEATH ROW DOG IS SAVED MINUTES FROM GAS CHAMBER—& BECOMES A SUPERSTAR! The bottom quarter announced BREAKTHROUGH PILL CURES ULCERS. Marshall's insights about ulcers shared the issue with other breakthrough stories, such as one on page 10: MY ZIPPER STOPPED ROBBER'S BULLET—AND SAVED MY LIFE. The story on the "breakthrough pill," at least, was done with a high degree of accuracy and with only a moderate dose of sensationalism. The report may have helped the anemic credibility of the *National Enquirer,* but the dependence of medical progress on that report does not inspire confidence in the portrayal of medicine as a science that deftly roots out truth from the evidence.

Such is the history of one of the great medical discoveries of the twentieth century. Why does the medical infrastructure fail to advance important biomedical discoveries more efficiently? The problem seems to be that the most important discoveries are often important because they change the established views, and those who control access to funding and the channels of scientific communication tend to be believers in the established views. Surprisingly and disappointingly, the importance of discoveries and their tendency to get lost in the structure of biomedical research thus arise from the same source: the discoveries are investigating a new angle that is at variance with established knowledge and therefore lacks credibility. Important papers on such new angles tend to get blasted by reviewers who see the problem only from their own angle. The papers might be rescued by editors with the broader vision that is needed to look beyond the criticism of the reviewers and the courage to withstand criticism from respected peers; but these assets appear to be rare in that risk-averse breed.

The lesson is timely because other chronic diseases—cancers, stroke, atherosclerosis, Alzheimer's, impotence—are now in a state of limbo very similar to the limbo of peptic ulcers during the 1980s and early 1990s. The same sort of dismissiveness of infectious causation that characterized the debate over peptic ulcers in the 1940s is pervasive in the current arguments about these chronic diseases.

One would hope and expect that prepared minds would no longer be the limited resource, but the track record of medicine does not offer reassurance. The lag in understanding peptic ulcers is a disturbing example, perhaps because it is so recent, but similar arguments could be made for a diverse group of ailments, including cervical cancer, atherosclerosis, and infertility. And these lags have been a repeating theme throughout the development of the germ theory of disease—they constitute the rule rather than the exception.

The germ theory itself is a case in point. Three centuries ago Antony van Leeuwenhoek made microscopic observations of "animalcules." If medical minds of his generation or of the several generations that followed had been prepared, these observations could have been linked to the theory of infectious causation that the Italian Girolamo Fracastoro had articulated a century and a half before, that Athanasius Kircher and Giovanni Maria Lancisi had defended during Leeuwenhoek's lifetime, and that Marcus Anton von Plenciz had supported in the century following Leeuwenhoek. A century and a half after Leeuwenhoek, Jacob Henle again proposed something that we can now recognize as a germ theory. Yet even in 1840, when Henle published his ideas, he was dismissed as "a gallant defender of an old-fashioned error." A few prepared minds were not enough. A critical mass of prepared minds was necessary to guide medicine to treat seriously the germ theory of disease, in spite of the logical consistency of the evidence. This critical mass was reached during the last quarter of the nineteenth century as Koch, Pasteur, and an army of microbe hunters developed such an overwhelming amount of evidence that even the most unprepared minds could grasp the validity of the theory for acute infectious diseases.

The same oversight has played out through the century and a half since Henle, up to—and including—the present. But now we are wrestling with the scope of chronic infectious diseases. Of course, anyone can point to the major exception: the golden era of the discovery of infectious causation as the nineteenth century turned into the twentieth. We can now recognize that the end of this golden era resulted not so much from the maturation of the germ theory as from its arrested development. The arrest has been and continues to be caused by unprepared minds with a myopic view of the future.

How then can minds become better prepared to recognize infectious causation? One step would be to recognize that the microbial world with which humans intimately interact is far more rich and diverse than we have thought and perhaps than we would like to think. A prepared mind recognizes that we know very little about what these microbes are doing. Three molecular biomedical scientists at Stanford recently demonstrated this point by conducting a study that would have made Leeuwenhoek chuckle with delight. Leeuwenhoek was not some snooty academic. He was an earthy and mischievous character who studied biology for personal enjoyment. He seemed to take delight in making seventeenth-century ladies squirm by showing them the microscopic worms that lived in vinegar. But he tempered his enthusiasm with a sense of etiquette: "They were keen on seeing the little eels in vinegar, but some of them were so disgusted at the spectacle that they vowed they'd never use vinegar again. But what if one should tell such people in future that there are more animals living on the scum on the teeth in a man's mouth than there are men in a whole kingdom, especially in those who don't ever clean their teeth. . . . I judge that all the people living in our united Netherlands are not as many as the living animals I carry in my mouth this very day."

Three hundred years later, Stanford scientists Ian Kroes, Paul Lepp, and David Relman picked up where Leeuwenhoek left off, using a molecular technique called polymerase chain reaction as a kind of molecular microscope to look at some scum from two teeth of one of the researchers. They probed the sample with a variety of DNA fragments, which could detect and amplify a great variety of possible DNA sequences. Their fishing expedition turned up evidence for more than thirty kinds of bacteria that were new to science. Their results, a shocker for people who smugly supposed that twentieth-century medicine had a good understanding of the intimate relationships between us and our microbes, also give a sense of how much work may be necessary to evaluate fully whether microbes are causally involved in disease. What are all those microbes doing in there? Might some of them be causing chronic disease? We don't know. But we do know that some inhabitants of the oral cavity cause tooth decay and gum disease. Some that were previously thought to be benign are now thought to be involved in heart disease and cancer. Presumably, many, if not most, of these organisms are obtained and have been maintained through countless generations by direct contact with things that are put in the mouth or by things that touch the mouth, perhaps through kissing. "To see what is in front of one's nose needs a constant struggle." George Orwell's dictum applies more broadly than he envisioned.

Orwell was referring to people who cling to a belief that they want to believe even though they are aware of contradictory information. As Orwell put it, people may hold a view that requires that they ignore "facts which are obvious and unalterable, and which will have to be faced sooner or later." Then, when the facts are faced, "one may simply forget that one ever held it." He illustrated his idea with sociopolitical examples, but his point applies just as forcefully to medicine. Orwell penned this social critique in 1946, when the infection and anti-infection experts were arguing over the cause of peptic ulcers. He died in 1950, when the anti-infection camp had won a victory that would last for more than a generation. Orwell's point applies as well to the anti-infection stance on human cancer during the half century after his death, a stance that is grudgingly being relinquished bit by bit in the face of accumulating evidence. How many of the older generation of medical leaders glibly dismiss infection as a cause of cancers of the breast, ovary, and prostate in the year 2000 but forget that they held similar views against the infectious causation of cervical cancer and liver cancer in 1975?

Orwell's point also applies to biomedical experts who claim that chronic diseases must be caused by bad genes, even when the genes would have to be so bad that they could not be maintained by a realistic rate of mutation. Minds need to be prepared to recognize when they are "ignoring facts which are obvious and unalterable." Prepared minds need to recognize that evolutionary processes are intimately involved in human disease. To deny this assertion is to deny the whole of biology. But integrating evolution into medicine requires a deep understanding of natural selection. To their credit, people in medicine have attempted to integrate evolutionary perspectives, but these attempts have often failed because of a superficial understanding of natural selection. These writers set themselves up for logical blunders and may confuse rather than enlighten their audience. The most common blunder is the presumption that natural selection favors stable harmony and long-term survival of host and parasite. Whether ecological harmony and long-term survival result from natural selection depends on whether these characteristics give the genes that encode them a competitive edge in the short run. If not, the genes and the characteristics will not persist over the long run.

One might hope that the basics and significance of natural selection would be embedded in nearly all well-educated minds before those minds encounter their first year of college, and certainly by the end of the last year. But this hope is far from being realized. The antievolution stance of some groups is partly responsible. Publishers of textbooks, worried about sales, have often quarantined their treatment of evolution to a few chapters or a few pages rather than giving evolution its rightful place as the fundamental unifying principle of biology. As such, it is more appropriately integrated into each chapter of a biology text to provide a sense of the "why" for each of the biological mechanisms discussed in most textbooks. The net effect is that most high school students are evolutionarily illiterate when they get their diplomas. They may get a bit of evolution in college, but as in the arts, math, and literature, expertise and profound insight are acquired only after immersion in the subject. College students go on to medical school, where they typically receive virtually no exposure to the evolutionary principles that unite biology. The newly minted M.D.'s believe that they have been packed full of the best biological insights that education can provide, when in fact they have often been deprived of biology's most fundamental insight. The intense specialization in Ph.D. programs leaves many newly minted Ph.D.'s in a similar situation. Biomedical researchers and medical practitioners thus often try to understand biological problems with a sorely inadequate understanding of the basic framework of biology.

The telltale signs of evolutionary illiteracy are not hard to spot. When one reads that disease organisms should always evolve to benign coexistence with their host, one is witnessing a telltale sign of evolutionary illiteracy. When one reads that harmful disease is a sign of recency in a host-parasite relationship, one is witnessing evolutionary illiteracy.

Though the progressive march of medicine is a common theme in medical texts, the truth is that medical progress has had a bungling sort of advancement. Rather than an army that has moved steadily toward the conquest of disease, it is more like a regiment that starts and stops and spins its wheels, and sometimes goes backward as it eventually moves forward along a muddy, rutted road. Consider any fundamental problem that has been decisively solved and trace the problem back through history, as I have just done for the germ theory of disease. Such detective work will usually reveal that the solutions could have been generated years, decades, and sometimes even centuries before they were actually put in place. More often than not the record will show that medical minds have not been adequately prepared to take advantage of the insights from basic science.

Even when the evidence is compelling, the resistance is great. Recall Edward Jenner's smallpox vaccine, the greatest breakthrough in the history of vaccination. As a member of the Royal Society of London, he submitted his paper to its journal for publication. The paper was flatly rejected, and Jenner was admonished: "A Fellow of the Society should be cautious and ought not to risk his reputation by presenting to the learned body anything which appears so much at variance with established knowledge, and withal so incredible." Jenner published the work on his own and then battled the medical dogma in England with evidence. It took a while. A decade after Jenner published his findings, well after the smallpox vaccine was used to great effect by Napoleon, the barbarians in the United States, and Jenner himself in England, parts of the medical establishment of England were still working hard to discredit it.

If one studies the history of hospital care and hospital-acquired infections, one encounters similar resistance to Ignaz Semmelweis, who spent his life demonstrating, in one of the pioneering efforts of epidemiology, that nineteenth-century hospital care was killing rather than helping mothers during delivery. For his insights, Semmelweis earned the ostracism of medicine. Only after the discovery of the causative *Streptococcus* by Pasteur could the unprepared minds of medicine grasp and accept the idea that invisible organisms were multiplying in people, causing disease and being transmitted by hospital attendants. Relying on logical argument more than exhaustive empirical study, Oliver Wendell Holmes Sr. came to the same conclusion about attendant-borne transmission even before Semmelweis. As a 34-year-old physician, he published his conclusions in the *New England Quarterly Journal of Medicine and Surgery* in 1843, four years before the 29-year-old Semmelweis first demonstrated transmission via the hands of the staff at the University of Vienna and eighteen years before Semmelweis published his massive collection of evidence. Like Semmelweis, he met violent opposition. In 1883, as Pasteur, Koch, and the other microbe hunters brought robust acceptance to the germ theory, the 74-year-old Holmes told a fellow physician, "I shrieked my warning louder and longer than any of them and I am pleased to remember that I took my ground on the existing evidence, before the little army of microbes was marched up to support my position."

It is not just acceptance of infectious causation; it is the history of medicine. Look backward or forward from the nineteenth century and you will find the same story. Consider the most demonstrable of medical sciences: anatomy. The pioneering work of Andreas Vesalius demonstrated that the Galenic dogma of the sixteenth century did not fit the facts. He wrote of his expectation that "old men filled with envy" would try to discredit his work. Vesalius knew what he was talking about. He was eventually pressured to resign his chair of anatomy at the University of Padua—one of the most prestigious positions in medicine at the time. He left after burning most of his writings, perhaps out of fear that they would be used against him.

We usually need decades of hindsight to have hard evidence of how medical progress has suffered from unprepared minds, but within that window there is evidence almost anywhere one cares to look. The delay in recognizing infectious causation of peptic ulcers is just the most recent demonstration; the historians of 2025 will be able to point to an array of new examples in the realms of cancer, cardiovascular disease, and mental illnesses.

So what will the decisive technologies of the future be? If diseases are caused by infection, the decisive technologies will probably be interventions that interfere with infection. Medicine generally tends to work under the assumption that the decisive technology needed to control a particular disease will be a modification of a decisive solution to some other disease. Vaccination against polio was a decisive technology borrowed from the decisive vaccines of the previous century and a half. Vaccination worked well for polio, but it holds much less promise as a decisive solution to such protean opponents as AIDS, malaria, and influenza. Indeed, the influenza vaccination program is probably best categorized as halfway technology. If medical researchers are not sufficiently imaginative and mentally prepared to envision new approaches and new perspectives, we will be left with the default outcome: some newly recognized diseases will be controlled by applications of the old approaches, but diseases that are resistant to the old approaches will persist and fester. This impasse characterizes attempts to use the successful vaccine programs against diseases like measles, diphtheria, and whooping cough as models for vaccination against AIDS, influenza, and malaria; and it characterizes the use of antibiotics as a model for antiviral drugs. No antiviral has ever earned the rank of decisive technology.

Though hard to predict, decisive technology is often easy to recognize after it has been enacted. The decisive effects of polio vaccines and penicillin, for example, were immediately apparent. But decisive technology is not always easy to recognize, and it can arise from unexpected places. Consider the vector-borne diseases and diarrheal diseases discussed in chapter 1. Decisive technology against these diseases was certainly enacted in North America and Europe throughout the twentieth century. We know that something knocked them out and keeps them knocked out in spite of a continual rain of the causative agents into these regions from surrounding areas. But public health experts do not agree on what that something is. It's not just the climate, because these areas supported a rich flora of these diseases a century ago. Some might point to mosquito abatement or the draining of swamps as the sole explanation for the failure of vector-borne diseases to become established. But mosquito control cannot be the only factor because there are plenty of places in the Euro-American theater where mosquito densities are very high, yet the malaria and dengue that continually slip across the border seem doomed to extinction. One decisive technology for malaria and dengue appears to be screening of houses, enhanced perhaps by air conditioning, television, and computer games. Though the vectors are still present in the face of this technology, their presence is insufficient to maintain the disease.

In the case of diarrheal diseases like cholera, dysentery, and typhoid, water purification seems to be the decisive technology. Here the appearance of agreement among public health experts is superficial. Typhoid, dysentery, and cholera disappeared in concert with the curbing of waterborne transmission. But experts tend to believe that water purification is not sufficient by itself because each of these diseases can be transmitted by other routes besides water. The conventional view is therefore that the whole spectrum of hygienic improvements, including sewage disposal, hand washing, improved sanitation in food preparation, and antibiotics, played a role. Evolutionary considerations, however, suggest that the provisioning of safe water supplies alone could have been the decisive technology. The bacteria that cause typhoid, dysentery, and cholera seem to have evolved to become so dependent on waterborne transmission that they cannot maintain themselves in their highly virulent states without it. For each of these diseases, the historical evidence supports this view because in each case we find milder forms maintaining themselves in areas with relatively safe water. Mild agents of dysentery maintain themselves in the United States, for example, even though the most dangerous killers cannot. Even a mild variant of the cholera organism can maintain itself along the coast of Louisiana and Texas, where protection of water and hygiene are substandard and where eating of raw shellfish appears to contribute to the maintenance of the organism. Yet Louisiana and Texas do not have a bad cholera problem. Fairly large numbers of people seem to become infected, but almost nobody gets cholera. Apparently the problem does not materialize because the endemic strains are benign. The correct explanation for the decisive technology has to be something that favored the mild strains over the harmful ones, and the leading candidate is the provisioning of safe water supplies.

• • •

The decisive technologies of the future will encompass both new applications of past solutions and new approaches. New applications of previous solutions will undoubtedly be decisive for some newly recognized infectious diseases without any need for new approaches; recent examples include hepatitis B, Lyme disease, and peptic ulcers. Other newly recognized infectious diseases will resemble diseases that have a long history of stymieing medicine, and they will continue to confound us unless we bring new approaches to bear on them; a recent example in this category is AIDS. This conclusion is based on the simple idea that the major diseases of the future will continue to share a basic similarity to the major diseases of the past that we now understand and sometimes have controlled—they will be caused largely by infection. By the luck of the draw, some of the causative agents will be vulnerable to recycled versions of decisive technologies used in the past, and others will not. The more important of the newly recognized infectious diseases will differ from the infectious diseases of the past primarily because they will be almost entirely chronic rather than acute. We already have the technological nuts and bolts needed to control most infectious diseases once we understand their infectious nature. Vaccines, antimicrobials, and hygienic improvements may control most heart disease, infertility, mental illnesses, and cancers, especially if these solutions are used not just to decimate pathogens but also to direct the evolution of the causative microbes.

If we lived in the small groups of our Stone Age ancestors but possessed the technology of twenty-first-century society, then perhaps we could eradicate many of these adversaries. Throughout most of the tenure of *Homo sapiens* the task would have been fairly manageable because the effective pool of pathogens that a person might encounter resided in perhaps just a few thousand people. But now the resource islands that our pathogens inhabit number over six billion, and residents on any one of these six billion mobile islands could travel by land, ocean, and air to contact almost any other island within a few weeks or at most a few years. For most of these pathogens, eradication will be hopeless for the same reason that surveillance is of limited value: there are too many interconnections in the global collection of six billion people. The spread, mixing, and remixing is already far beyond what could be contained. The dike has thousands of holes, and the boy has only two thumbs. It is time to step back and prepare our minds to recognize the solutions that are in front of our noses.

NOTES

Chapter 1: The Virulence of Acute Infections

11 *leading authorities in the health sciences believed:* T. Smith, *Parasitism and Disease* (Princeton, N.J.: Princeton University Press, 1934); R. Dubos, *Man Adapting* (New Haven, Conn.: Yale University Press, 1965); L. Thomas, "Notes of a biology-watcher: Germs," *New England Journal of Medicine* 247 (1972): 553–5.

13 *diseases transmitted by mosquitoes, tsetse flies, lice, and sandflies:* P. W. Ewald, "Host-parasite relations, vectors, and the evolution of disease severity," *Annual Review of Ecology & Systematics* 14 (1983): 465–85.

13 *an account by the tropical-disease expert Alan Spira:* A. Spira, "Dengue Fever," February 13, 1998; available at <www.armchair.com/info/spira7.html>.

15 *the more waterborne the diarrheal bacterium, the more deadly it is:* P. W. Ewald, "Waterborne transmission and the evolution of virulence among gastrointestinal bacteria," *Epidemiology and Infection* 106 (1991): 83–119.

16 *the same disease that physicians and medical students had been studying in the morgue:* L. F. Céline, *Mea Culpa* and *The Life and Work of Semmelweis,* trans. R. A. Parker (Boston: Little, Brown & Co., 1937).

18 *J. F. von Arneth, arrived in Edinburgh to present Semmelweis's findings:* F. G. Slaughter, *Immortal Magyar: Semmelweis, Conqueror of Childbed Fever* (New York: Henry Schuman, 1950), pp. 138–9.

18 *Lister remembered that sewage had been treated:* C. E. A. Winslow, *The Conquest of Epidemic Disease* (Princeton, N.J.: Princeton University Press, 1943), pp. 301–2.

19 *the tenth leading cause of death:* R. W. Haley et al., "The nationwide nosocomial infection rate: A new need for vital statistics," *American Journal of Epidemiology* 121 (2) (1985): 159–67; M. C. White, "Mortality associated with nosocomial infections—analysis of multiple cause-of-death data," *Journal of Clinical Epidemiology* 46 (1993): 95–100.

19 *a study of a nursery ward for newborns:* T.N.K. Raju and D. Kobler, "Improving handwashing habits in the newborn nurseries," *American Journal of Medical Science* 302 (1991): 355–8.

20 *when antibiotic-resistant and antibiotic-sensitive hospital strains are compared:* P. W. Ewald, "Guarding against the most dangerous emerging pathogens: Insights from evolutionary biology," *Emerging Infectious Diseases* 2 (1996): 245–57.

20 Escherichia coli: P. W. Ewald, "Transmission modes and the evolution of virulence, with special reference to cholera, influenza and AIDS," *Human Nature* 2 (1991): 1–30.

24 *one of the twentieth century's leading experts on infectious disease, Macfarlane Burnet, tried to reconstruct:* F. M. Burnet and E. Clark, *Influenza: A Survey of the Last 50 Years in the Light of Modern Work on the Virus of Epidemic Influenza* (Melbourne, Australia: Macmillan, 1942).

26 *smallpox scabs:* H. L. Wolff and J.J.A.B. Croon, "The survival of smallpox virus (*Variola minor*) in natural circumstances," *Bulletin of the World Health Organization* 38 (1968): 492–3.

26 *smallpox viruses that were lying in wait:* J. A. Poupard, L. A. Miller, and L. Granshaw, "A new look at the smallpox story," *Amherst,* winter 1989: 18–31. (Reprinted from *ASM News,* March 1989, American Society for Microbiology.)

28 *The rabbits were released in 1859:* F. M. Burnet and D. O. White, *Natural History of Infectious Disease* (Cambridge: Cambridge University Press, 1972), p. 139.

29 *the virus was first introduced in the 1950s:* F. Fenner and F. N. Ratcliffe, *Myxomatosis* (London: Cambridge University Press, 1965).

Chapter 2: The Long Fuse of Sexually Transmitted Diseases

39 *Japanese Americans and Caribbean Americans:* P. H. Levine et al., "The effect of ethnic differences on the pattern of HTLV-I-associated T-cell leukemia/lymphoma (HATL) in the United States," *International Journal of Cancer* 56 (1994): 177–81.

39 *Genital herpes simplex viruses:* K. Sunagawa et al., "Pathologic studies and comparison of the virulence of herpes simplex virus type 2 from Okinawa, Japan, and Chiang Mai, Thailand," *International Journal of Experimental Pathology* 76 (4) (1995): 255–62.

39 *women who have more sexual partners:* E. L. Franco et al., "Transmission of cervical human papillomavirus infection by sexual activity: Differences be-

tween low and high oncogenic risk types," *Journal of Infectious Diseases* 172 (1994): 756–63.

39 *During the war in the former Yugoslavia:* M. Grce et al., "Increase of genital human papillomavirus infection among men and women in Croatia," *Anticancer Research* 16 (2) (1996): 1,039–41.

40 *prostitutes in Fukuoka City, Japan:* M. Tanaka et al., "Trends in sexually transmitted diseases in Fukuoka City, Japan, 1990–93," *Genitourinary Medicine* 72 (5) (1996): 358–61.

40 *to harm hosts perpetually:* G. M. Cochran, P. W. Ewald, and K. D. Cochran, "Infectious causation of disease: An evolutionary perspective," *Perspectives in Biology and Medicine* 43 (2000): 406–48.

Chapter 3: The Stealth of the Chronic

45 A Dictionary of Medicine: R. Quainn, *A Dictionary of Medicine* (New York: Appleton, 1884).

47 *the French government sent Nicholas Chervin:* W. H. McNeill, *Plagues and Peoples* (Garden City, N.J.: Anchor, 1976).

47 *no trace is left behind:* C.E.A. Winslow, *The Conquest of Epidemic Disease* (Princeton, N.J.: Princeton University Press, 1943).

49 *shingles is a delayed manifestation of chicken pox:* J. von Bókay, "Ueber den ätiologischen Zusammenhang der Varizellen mit gewissen Fällen von Herpes zoster," *Wiener Klinische Wochenschrift* 22 (1909): 1,323–6.

49 *rheumatic fever:* G. H. Stollerman, *Rheumatic Fever and Streptococcal Infection* (New York: Grune & Stratton, 1975), p. 16.

53 *U.S. surgeon general William H. Stewart:* W. H. Stewart, "A mandate for state action," presented at the Association of State and Territorial Health Officers, Washington, D.C., December 4, 1967.

55 *cancers that appear to be on the verge of being ascribed to infectious causation:* M. L. Labat, "Possible retroviral etiology of human breast cancer," *Biomedicine & Pharmacotherapy* 52 (1998): 6–12; B.G.T. Pogo and J. F. Holland, "Possibilities of a viral etiology for human breast cancer—a review," *Biological Trace Element Research* 56 (1997): 131–42; E. M. Rakowicz-Szulczynska, B. Jackson, and W. Snyder, "Prostate, breast and gynecological cancer markers RAK with homology to HIV-1," *Cancer Letters* 124 (1998): 213–23; M. Bonnet et al., "Detection of Epstein-Barr virus in invasive breast cancers," *Journal of the National Cancer Institute* 91 (16) (1999): 1,376–81; J. Parsonnet, *Microbes and Malignancy: Infections As a Cause of Human Cancer* (New York: Oxford, 1999).

55 *genetic causation:* G. M. Cochran, P. W. Ewald, and K. D. Cochran, "Infectious causation of disease: An evolutionary perspective," *Perspectives in Biology and Medicine* 43 (2000): 406–48.

Chapter 5: The Endless War

71 *U.S. political leaders of the 1950s rallied the population:* Laurie Garrett presents the social and political setting for these statements and traces them to their sources in chapter 2 of her book *The Coming Plague* (New York: Farrar, Straus and Giroux, 1994).

72 *"the future of infectious diseases will be very dull"*: F. M. Burnet and D. O. White, *Natural History of Infectious Disease* (Cambridge: Cambridge University Press, 1972).

72 The Conquest of Disease: C.E.A. Winslow, *The Conquest of Epidemic Disease* (Princeton, N.J.: Princeton University Press, 1943).

73 *John W. Huggins:* W. Orent, "Killer pox in the Congo," *Discover* 20 (10) (1999): 74–9.

74 *Isao Arita:* L. Garrett, *The Coming Plague* (New York: Farrar, Straus and Giroux, 1994).

76 *Michele Carbone:* M. Carbone et al., "Simian virus 40–like DNA sequences in human pleural mesothelioma," *Oncogene* 9 (1994): 1,781–90

76 *monkey virus:* M. Carbone et al., "New molecular and epidemiological issues in mesothelioma: Role of SV40," *Journal of Cellular Physiology* 180 (1999): 167–72.

Chapter 6: Where Are They Coming From?

83 *a melting pot for the world's pathogens:* J. Diamond, *Guns, Germs and Steel* (New York: W. W. Norton, 1997).

83 *the West Nile virus:* S. Milius, "Animal whodunit, medical mystery: Scientists cross species barriers to diagnose West Nile encephalitis," *Science News* 156 (1999): 378–80.

87 *The dengue virus is continually slipping across the Texas border:* J. Gill, L. M. Stark, and G. G. Clark, "Dengue surveillance in Florida, 1997–98," *Emerging Infectious Diseases* 6 (2000): 30–5.

88 *as with dengue, the chain of infection petered out:* J. B. Weissman et al., "Impact in the U.S. of the Shiga dysentery pandemic of Central America and Mexico:

A review of surveillance data through 1972," *Journal of Infectious Diseases* 129 (1974): 218–23.

88 *Remembering the nineteenth-century experience:* G. García Márquez, *El amor en los tiempos del cólera* (translated into English as *Love in the Time of Cholera*) (Bogotá, Colombia: Editorial Oveja Negra, 1985).

89 *It entered Latin America by the back door:* R. V. Tauxe, E. D. Mintz, and R. E. Quick, "Epidemic cholera in the new world: Translating field epidemiology into new prevention strategies," *Emerging Infectious Diseases* 1 (1995): 141–6.

Chapter 7: Malignant Growths in Our Backyard

89 *These proteins neutralize our bodies' safeguards:* H. zur Hausen, "Viruses in human cancers," *Science* 254 (1991): 1,167–73; R.J.C. Slebos et al., "Functional consequences of directed mutations in human papillomavirus E6 proteins: Abrogation of p53-mediated cell cycle arrest correlates with p53 binding and degradation in vitro," *Virology* 208 (1995): 111–20.

89 *E6 also interferes with the cell's self-destruct mechanism:* P. J. Duerksen-Hughes, J. Yang, and S. B. Schwartz, "HPV 16 E6 blocks TNF-mediated apoptosis in mouse fibroblast LM cells," *Virology* 264 (1999): 55–65.

97 *lethal types outnumbered the more benign types:* M. Grce et al., "Increase of genital human papillomavirus infection among men and women in Croatia," *Anticancer Research* 16 (1996): 1,039–41.

97 *By the third year after the war, the lethal types had begun to recede:* M. Grce et al., "Detection and typing of human papillomaviruses by polymerase chain reaction in cervical scrapes of Croatian women with abnormal cytology," *European Journal of Epidemiology* 13 (1997): 645–51.

97 *the risks of acquiring the harmful and benign viruses:* E. L. Franco et al., "Transmission of cervical human papillomavirus infection by sexual activity: Differences between low and high oncogenic risk types," *Journal of Infectious Diseases* 172 (1994): 756–63.

98 *searching for an agent of Kaposi's sarcoma:* P. S. Moore and Y. Chang, "Kaposi's sarcoma (KS), KS-associated herpesvirus, and the criteria for causality in the age of molecular biology," *American Journal of Epidemiology* 147 (1998): 217–21.

99 *Arthur Boettcher:* A. Boettcher, "Zur Genese des perforirenden Magengeschwürs," *Dorpater Medicinische Zeitschrift* 5 (1874): 148–51.

99 *experimentally transmitting the bacterium:* J. L. Doenges, "Spirochetes in gastric glands of *Macacus rhesus* and humans without definite history of related disease," *Proceedings of the Society for Experimental Biology and Medicine* 38

(1938): 536–8; M. Kidd and I. M. Modlin, "A century of Helicobacter pylori: Paradigms lost—paradigms regained," *Digestion* 59 (1998): 1–15.

99 *"Use it. It works":* P. Fremont-Smith, "Letter to the editor," *Atlantic Monthly* 283 (5) (1999): 12.

99 *attributed peptic ulcers to . . . everything but infection:* M. I. Grossman, "Peptic ulcer: Pathogenesis and pathophysiology," in *Cecil Textbook of Medicine,* 15th ed., ed. P. B. Beeson, W. McDermott, and J. B. Wyngaarden (Philadelphia: Saunders, 1979), pp. 1502–7; S. Goodwin, "Historical and microbiological perspectives," in *Helicobacter pylori Infection: Pathophysiology, Epidemiology, and Management,* ed. T. C. Northfield, M. Mendall, and P. M. Goggin (Deventer, Netherlands: Kluwer, 1993), pp. 1–9.

103 *Human parvovirus B19:* I. Nakashima, K. Fujihara, and Y. Itoyama, "Human parvovirus B19 infection in multiple sclerosis," *European Neurology* 42 (1999): 36–40.

103 *Almost 40 percent of the clinic patients:* L. Salvatori et al., "Seroprevalence of anti-human parvovirus B19 antibodies in patients attending a centre for sexually transmitted diseases," *Microbiologica* 22 (1999): 181–6.

105 *evidence of retroviruses in breast tumors:* B.G.T. Pogo and J. F. Holland, "Possibilities of a viral etiology for human breast cancer—a review," *Biological Trace Element Research* 56 (1997): 131–42; E. M. Rakowicz-Szulczynska, B. Jackson, W. Snyder, "Prostate, breast and gynecological cancer markers RAK with homology to HIV-1," *Cancer Letters* 124 (1998): 213–23.

105 *mouse mammary tumor virus:* Y. Wang et al., "Expression of mammary tumor virus-like env gene sequences in human breast cancer," *Clinical Cancer Research* 4(2)(1998): 565–8.

105 *More than one third of the human breast cancers:* Y. Wang et al., "Detection of mammary tumor virus env gene-like sequences in human breast cancer," *Cancer Research* 55 (1995): 5,173–9.

105 *The geographic distribution of breast cancer:* T. H. Stewart et al., "Breast cancer incidence highest in the range of one species of house mouse, *Mus domesticus,*" *British Journal of Cancer* 82 (2000): 446–51.

105 *Epstein-Barr virus:* M. Bonnet et al., "Detection of Epstein-Barr virus in invasive breast cancers," *Journal of the National Cancer Institute* 91 (1999): 1,376–81.

105 *a study of women in Wales and England:* A. J. Swerdlow et al., "Risks of breast and testicular cancers in young adult twins in England and Wales: Evidence on prenatal and genetic aetiology," *Lancet* 350 (1997): 1,723–8.

Chapter 8: Our Vulnerable Hearts and Minds

108 *they published their findings in the* New England Journal of Medicine: J. T. Grayston et al., "A new *Chlamydia psittaci* strain, TWAR, isolated in acute respiratory tract infections," *New England Journal of Medicine* 315 (1986): 161–8.

108 *over half of them showed signs of infection:* J. T. Grayston et al., "Community and hospital acquired pneumonia associated with *Chlamydia* TWAR infection demonstrated serologically," *Archives of Internal Medicine* 149 (1989): 169–73; P. Saikku, "The epidemiology and significance of *Chlamydia pneumoniae*," *Journal of Infection* 25 (1992): 27–34.

109 *the Seattle researchers named the bacterium:* J. T. Grayston et al., "*Chlamydia pneumoniae* sp. nov. for *Chlamydia* sp. strain TWAR," *International Journal of Systematic Bacteriology* 39 (1989): 88–90.

109 *heart attack patients had antibodies:* P. Saikku, "*Chlamydia pneumoniae* infection as a risk factor in acute myocardial infarction," *European Heart Journal* 14 (supplement K) (1993): 62–5.

111 *confirm definitively the presence of the organism:* C.-C. Kuo and L. A. Campbell, "Is infection with *Chlamydia pneumoniae* a causative agent in atherosclerosis?" *Molecular Medicine Today,* October 1998: 426–30.

112 *a story on atherosclerosis for* Forbes *magazine:* P. Ross, "Do Germs Cause Cancer?" *Forbes,* November 15, 1999, 194–200.

112 *C-reactive protein:* A. J. Szalai et al., "C-reactive protein: Structural biology and host defense function," *Clinical Chemistry and Laboratory Medicine* 37 (1999): 265–70.

113 *When was the idea of infectious causation of atherosclerosis dropped?:* F. J. Nieto, "Infections and atherosclerosis: New clues from an old hypothesis?" *American Journal of Epidemiology* 148 (1998): 937–48.

114 C. pneumoniae *lives within our cells:* P. Saikku, "Chronic *Chlamydia pneumoniae* infections," in *Chlamydia Pneumoniae Infection,* ed. L. Allegra and F. Blasi (Berlin: Springer, 1995).

117 *The last category was a big one:* C. T. Williams, "Phthisis," in *A Dictionary of Medicine,* ed. R. Quainn (New York: Appleton, 1884), 1,166–83.

118 *gastric acidity:* M. I. Grossman, "Peptic ulcer: Pathogenesis and pathophysiology," in *Cecil Textbook of Medicine,* 15th ed., ed. P. B. Beeson, W. McDermott, and J. B. Wyngaarden (Philadelphia: Saunders, 1979), 1,502–7.

118 *hyperpepsinogenemia:* P. Walker et al., "Life events stress and psychosocial factors in men with peptic ulcer disease. II. Relationships with serum pepsino-

gen concentrations and behavioral risk factors," *Gastroenterology* 94 (1988): 323–30.

118 *The ϵ4 allele has been identified as a risk factor for atherosclerosis and stroke:* Y. Ji et al., "Apolipoprotein E polymorphism in patients with Alzheimer's - disease, vascular dementia and ischemic cerebrovascular disease," *Dementia and Geriatric Cognitive Disorders* 9 (1998): 243–5; E. Ilveskoski et al., "Age-dependent association of apolipoprotein E genotype with coronary and aortic atherosclerosis in middle-aged men—an autopsy study," *Circulation* 100 (1999): 608–13; R. W. Mahley and Y. D. Huang, "Apolipoprotein E: From atherosclerosis to Alzheimer's disease and beyond," *Current Opinion in Lipidology* 10 (1999): 207–17.

119 *Some clever research on arthritis patients:* H. C. Gérard et al., "Frequency of apolipoprotein E (APOE) allele types in patients with *Chlamydia*-associated arthritis and other arthritides," *Microbial Pathogenesis* 26 (1998): 35–43.

120 *High iron levels are an environmental risk factor for atherosclerosis:* B. deValk and J.J.M. Marx, "Iron, atherosclerosis, and ischemic heart disease," *Archives of Internal Medicine* 159 (1999): 1,542–8.

120 *high iron levels:* J. L. Sullivan and E. D. Weinberg, "Iron and the role of *Chlamydia pneumoniae* in heart disease," letter in *Emerging Infectious Diseases* 5 (1999): 724–6.

121 *The other risk factors are similarly better explained:* M. Leinonen and P. Saikku, "Interaction of *Chlamydia pneumoniae* infection with other risk factors of atherosclerosis," *American Heart Journal* 138 (1999): S504–6.

122 *There have now been about ten studies:* R. J. Genco et al., "Cardiovascular diseases and oral infections," in *Periodontal Medicine,* ed. L. F. Rose et al. (Lewiston, N.Y.: B. C. Decker, 1999), 62–82; T. Wu et al., "Periodontal disease and cerebrovascular disease: The first national health and nutrition examination survey and its follow-up study," *Archives of Internal Medicine* 2000: (in press).

125 *brains from the Alzheimer's patients were positive for* C. pneumoniae: B. J. Balin et al., "Identification and localization of *Chlamydia pneumoniae* in the Alzheimer's brain," *Medical Microbiology and Immunology* 187 (1998): 23–42.

Chapter 9: Diseases of Blood and Steel

127 The River: E. Hooper, *The River: A Journey to the Source of HIV and AIDS* (Boston: Little, Brown & Co., 1999).

130 *Walter Kyle:* W. S. Kyle, "Simian retroviruses, poliovaccine, and the origin of AIDS," *Lancet* 339 (1992): 600–1.

132 *If that diversification occurred in humans:* B. H. Hahn et al., "AIDS as a zoonosis: Scientific and public health implications," *Science* 287 (2000): 607–14.

132 *two fascinating viruses:* J. Takehisa et al., "Natural infection of chimpanzees with new lentiviruses related to HIV-1/SIVcpz," *Virology* 28 (1999): 169–73.

133 *One group of viruses:* I. Mboudjeka et al., "HIV type 1 genetic variability in the northern part of Cameroon," *AIDS Research and Human Retroviruses* 15 (1999): 951–6.

133 *other group M subtypes are known to recombine:* M. Peeters et al., "Characterization of a highly replicative intergroup M/O human immunodeficiency virus type 1 recombinant isolated from a Cameroonian patient," *Journal of Virology* 73 (1999): 7,368–75; J. Takehisa et al., "Human immunodeficiency virus type 1 intergroup (M/O) recombination in Cameroon," *Journal of Virology* 73 (1999): 6,810–20.

135 *infections in Senegal best meet this criterion:* P. W. Ewald, *Evolution of Infectious Disease* (New York: Oxford University Press, 1994).

135 *Most people who are infected with HIV-2 in Senegal:* P. J. Kanki, "Virologic and biologic features of HIV-2," in *AIDS and Other Manifestations of HIV Infection,* 2d ed., ed. G. P. Wormser (New York: Raven Press, 1992), 85–93.

135 *HIV-2 infections during the late 1980s:* P. Kanki et al., "Prevalence and risk determinants of human immunodeficiency virus type 2 (HIV-2) and human immunodeficiency virus type 1 (HIV-1) in west African female prostitutes," *American Journal of Epidemiology* 136 (1992): 895–907.

135 *data from Senegal:* Ibid.

136 *the diversity of their genetic sequences is comparable:* J. C. Hunt et al., "Envelope sequence variability and serologic characterization of HIV type 1 group O isolates from Equatorial Guinea," *AIDS Research and Human Retroviruses* 13 (1997): 995–1,005; J. C. Hunt et al., "Molecular analysis of HIV group O and HIV-2 variants from Africa," *Leukemia* 13 (supplement 3) (1997): 138–41; M. E. Quinones-Mateu et al., "Analysis of pol gene heterogeneity, viral qualispecies, and drug resistance in individuals infected with group O strains of human immunodeficiency virus type 1," *Journal of Virology* 72 (1998): 9,002–15; W. Janssens et al., "Interpatient genetic variability of HIV-1 group O," *AIDS* 13 (1999): 41–8; A. Mas et al., "Phylogeny of HIV type 1 group O isolates based on env gene sequences," *AIDS Research and Human Retroviruses* 15 (1999): 769–73.

144 *The rate of viral evolution in intravenous drug users:* M. Salemi et al., "Different population dynamics of human T-cell lymphotropic virus type II in intravenous drug users compared with endemically infected tribes," *Proceedings of the National Academy of Sciences of the United States of America* 96 (1999): 13,253–8.

144 *Retrospective studies of blood:* L. Garrett, *The Coming Plague* (New York: Farrar, Straus and Giroux, 1994), p. 363.

145 *yellow fever vaccination program:* L. B. Seeff et al., "A serologic follow-up of the 1942 epidemic of post-vaccination hepatitis in the United States Army," *New England Journal of Medicine* 316 (1987): 965–70.

145 *dangerous lung cancer:* G. R. Stenton, "Asbestos, simian virus 40 and malignant mesothelioma," *Thorax* 52 (supplement 3) (1997): S52–7.

145 *The virus sabotages:* M. Carbone et al., "New molecular and epidemiological issues in mesothelioma: Role of SV40," *Journal of Cellular Physiology* 180 (1999): 167–72.

146 *SV40 is not just an innocent bystander:* S. S. Murthy and J. R. Testa, "Asbestos, chromosomal deletions, and tumor suppressor gene alterations in human malignant mesothelioma," *Journal of Cellular Physiology* 180 (1999): 150–7.

146 *Whether it causes brain tumors:* H. N. Zhen et al., "Expression of the simian virus 40 large tumor antigen (Tag) and formation of Tag-p53 and Tag-pRb complexes in human brain tumors," *Cancer* 86 (1999): 2,124–32.

146 *The most recent studies:* S. G. Fisher, L. Weber, and M. Carbone, "Cancer risk associated with simian virus 40 contaminated polio vaccine," *Anticancer Research* 19 (1999): 2,173–80.

146 *It has been recovered:* S. Jafar et al., "Serological evidence of SV40 infections in HIV-infected and HIV-negative adults," *Journal of Medical Virology* 54 (1998): 276–84; J. S. Butel et al., "Molecular evidence of simian virus 40 infections in children," *Journal of Infectious Diseases* 180 (1999): 884–7.

146 *The source of these SV40 infections:* A. R. Stewart, T. A. Lednicky, and J. S. Butel, "Sequence analyses of human tumor-associated SV40 DNAs and SV40 viral isolates from monkeys and humans," *Journal of Neurovirology* 4 (1998): 182–93.

146 *linked to colon cancer:* L. Laghi et al., "JC virus DNA is present in the mucosa of the human colon and in colorectal cancers," *Proceedings of the National Academy of Sciences of the United States of America* 96 (1999): 7,484–9; J. V. Neel, "The Colonel Harlan D. Sanders Award Address for 1998: JC virus and its possible role in oncogenesis," *American Journal of Medical Genetics* 83 (1999): 152–6.

Chapter 10: Modern Miasmas

150 *ancient Egyptian physicians purged intestinal worms:* I. H. Slater, "Strychnine, picrotoxin, pentylenetetrazol, and miscellaneous drugs," in *Drill's Pharmacology in Medicine*, 3d ed., ed. J. R. DiPalma (New York: Blakiston/McGraw-Hill, 1965), pp. 379–93.

152 *Infectious causation of cancer:* P. A. Rous, "Sarcoma of the fowl transmissible by an agent separable from the tumor cells," *Journal of Experimental Medicine* 13 (1911): 397–411.

155 *Evidence of infection with Borna disease virus:* D. E. Dietrich et al., "A viro-psycho-immunological disease-model of a subtype affective disorder," *Pharmacopsychiatry* 31 (1998): 77–82; K. Iwahashi et al., "Positive and negative syndromes, and Borna disease virus infection in schizophrenia," *Neuropsychobiology* 37 (1998): 59–64.

155 *in these two mental illnesses:* M.V. Pletnikov et al., "Persistent neonatal Borna disease virus (BDV) infection of the brain causes chronic emotional abnormalities in adult rats," *Physiology & Behavior* 66 (1999): 823–31.

155 *schizophrenia expert Fuller Torrey:* E. F. Torrey, personal communication.

156 *schizophrenia is also more common:* E. F. Torrey et al., "Seasonality of births in schizophrenia and bipolar disorder: A review of the literature," *Schizophrenia Bulletin* 28 (1997): 1–38; E. F. Torrey, A. E. Bowler, and K. Clark, "Urban birth and residence as risk factors for psychoses: An analysis of 1880 data," *Schizophrenia Research* 25 (1997): 169–76.

158 *All the symptoms that are seen in patients who are ill:* R. Nesse, "Is depression an adaptation?" *Archives of General Psychiatry* 57 (2000): 14–20.

158 *A circuit may generate a particular sensation:* J. Tooby and L. Cosmides, "The past explains the present: Emotional adaptations and the structure of ancestral environments," *Ethology and Sociobiology* 11 (1990): 375–424.

158 *a swelling in the basal ganglia:* J. N. Giedd et al., "MRI assessment of children with obsessive-compulsive disorder or tics associated with streptococcal infection," *American Journal of Psychiatry* 157(2) (2000): 281–3.

158 *as association with streptococcal infection:* J. L. Rapoport, "The new biology of obsessive-compulsive disease: Implications for evolutionary psychology," *Perspectives in Biology and Medicine* 41 (1998): 159–75.

162 *Japanese researchers:* T. Kitani et al., "Possible correlation between Borna disease virus infection and Japanese patients with chronic fatigue syndrome," *Microbiology and Immunology* 40 (7) (1996): 459–62; T. Nakaya et al., "Demonstration of Borna disease virus RNA in peripheral blood mononuclear cells derived from Japanese patients with chronic fatigue syndrome," *FEBS Letters* 378 (2) (1996): 145–9.

Chapter 11: Reverberations Across Society

167 *Vaccine Adverse Event Reporting System:* J. A. Singleton et al., "An overview of the vaccine adverse event reporting system (VAERS) as a surveillance system," *Vaccine* 17 (1999): 2,908–17.

169 *the case of Keil Chemical:* J. Morris, "Small-time polluter, big-time problems," *U.S. News & World Report* 128 (8) (2000): 57–8.

174 *Depressive illnesses:* K. R. Jamison, *Touched with Fire* (New York: The Free Press, 1993).

Chapter 12: Biobombs

177 *"a humble cottage in the country":* H. G. Unger, *Noah Webster: The Life and Times of an American Patriot* (New York: John Wiley, 1998).

179 *"Inoculate the Indians":* J. A. Poupard, L. A. Miller, and L. Granshaw, "A new look at the smallpox story," *Amherst,* winter 1989, 18–31 (reprinted from the American Society for Microbiology's *ASM News,* March 1989).

179 *a strange "German disease":* A. Beevor, *Stalingrad* (New York: Viking, 1998).

180 *The professor then told Alibek:* K. Alibek, *Biohazard* (New York: Random House, 1999), p. 30.

Chapter 13: The Protean Opponents

197 Malaria: Waiting for the Vaccine: G. A. Targett, ed., *Malaria: Waiting for the Vaccine* (Chichester, England: John Wiley and Sons, 1991).

197 *Kamini Mendis:* K. N. Mendis, "Malaria vaccine research—a game of chess," in *Malaria: Waiting for the Vaccine,* ed. G. A. Targett (Chichester, England: John Wiley and Sons, 1991), 183–96.

201 *the prevalence was less than 1 percent:* R. B. Watson, "Location and mosquito-proofing of dwellings," in *Malariology: A Comprehensive Survey of All Aspects of This Group of Diseases from a Global Standpoint,* vol. 2, ed. M. F. Boyd (Philadelphia: Saunders, 1949), pp. 1,184–202.

208 *I first wrote about this general prediction in 1993:* P. W. Ewald, *Evolution of Infectious Disease* (New York: Oxford University Press, 1994).

208 *in the context of Senegal:* P. W. Ewald, "Pedigree of a retrovirus," review of *Viral Sex: The Nature of AIDS,* by J. Goudsmit, *Natural History* 106 (5) (1997): 8–9.

208 *HIV-1 has* not *displaced HIV-2 in Senegal:* N. Meda et al., "Low and stable HIV infection rates in Senegal: Natural course of the epidemic or evidence for success of prevention?" *AIDS* 13 (1999): 1,397–405.

Chapter 14: Tools of Domestication

216 *the organism is still present and transmissible:* Q. S. He et al., "Outcomes of *Bordetella* infections in vaccinated children: Effects of bacterial number in

the nasopharynx and patient age," *Clinical and Diagnostic Laboratory Immunology* 6 (1999): 534–6.

218 *cells that were engineered to express these proteins were destroyed:* G. B. Lipford et al., "Peptide engineering allows cytotoxic T-cell vaccination against human papilloma virus tumour antigen, E6," *Immunology* 84 (1995): 298–303; M. E. Ressing et al., "Human CTL epitopes encoded by human papillomavirus type 16 E6 and E7 identified through *in vivo* and *in vitro* immunogenicity studies of HLA-a*0201-binding peptides," *Journal of Immunology* 154 (1995): 5,934–43.

223 *Few children show immunological signs of infection:* J. T. Grayston, "Infections caused by *Chlamydia pneumoniae* strain TWAR," *Clinical Infectious Diseases* 15 (1992): 757–63; P. Saikku, "The epidemiology and significance of *Chlamydia pneumoniae*," *Journal of Infection* 25 (1992): 27–34; S. Einarsson et al., "Age specific prevalence of antibodies against *Chlamydia pneumoniae* in Iceland," *Scandinavian Journal of Infectious Diseases* 26 (1994): 393–7; A. L. Koivisto et al., "Chlamydial antibodies in an elderly Finnish population," *Scandinavian Journal of Infectious Diseases* 31 (1999): 135–9.

223 *About one third of the parents:* R. W. Wissler and J. P. Strong, "Risk factors and progression of atherosclerosis in youth," *American Journal of Pathology* 153 (1998): 1,023–33.

Chapter 15: The Prepared Mind

227 *a little essay packed with insight:* L. Thomas, "The technology of medicine," *New England Journal of Medicine* 285 (1971): 1,366–8.

229 *Heinlein wrote about space shuttles:* R. A. Heinlein, *The Green Hills of Earth* (New York: Signet, 1951).

229 *a character trying to fix a computer:* W. M. Miller, *Canticle for Liebowitz* (Philadelphia: Lippincott, 1959).

229 *people using nuclear engines the size of a peanut:* I. Asimov, *The Second Foundation* (Garden City, N.Y.: Doubleday, 1953).

230 *gave their society a pocket computer:* L. Niven and J. Pournelle, *The Mote in God's Eye* (New York: Simon and Schuster, 1974).

234 *in the century following Leeuwenhoek:* C.E.A. Winslow, *The Conquest of Epidemic Disease* (Princeton, N.J.: Princeton University Press, 1943).

234 *even in 1840, when Henle published his ideas:* E. H. Ackerknecht, *A Short History of Medicine* (New York: Ronald Press, 1955).

235 *Three molecular biomedical scientists at Stanford:* I. Kroes, P. W. Lepp, and D. A. Relman, "Bacterial diversity within the human subgingival crevice," *Pro-*

ceedings of the National Academy of Sciences of the United States of America 96 (1999): 14,547–52.

236 “*To see what is in front of one’s nose*” and “facts which are obvious and unalterable”: G. Orwell, “In front of your nose,” in *In Front of Your Nose, 1945–1950,* vol. 4 of *The Collected Essays, Journalism and Letters of George Orwell,* ed. S. Orwell and I. Angus (New York: Harcourt Brace, 1968), pp. 122–5. (This article was originally published in the *Tribune,* 22 March 1946.)

240 *the 74-year-old Holmes told a fellow physician:* F. G. Slaughter, *Immortal Magyar: Semmelweis, Conqueror of Childbed Fever* (New York: Henry Schuman, 1950), p. 203.

GLOSSARY

acute infectious diseases Infectious diseases characterized by a distinct onset soon—usually within a few days to two weeks—after infection. Infectious causation was accepted by the end of the first decade of the twentieth century for most acute infectious diseases of humans.

AIDS Abbreviation for acquired immuno deficiency syndrome, which is an umbrella term for a variety of life-threatening diseases associated with HIV-induced decimation of the immune system.

allele One of the alternative genetic instructions for a gene.

antibodies Protein molecules that are released from B lymphocytes and bind to antigens, which may protrude from the surfaces of pathogens. This binding facilitates recognition and destruction of the pathogens by other cells in the immune system.

antigens Compounds that trigger the production of antibodies by the host and bind to the antibodies produced. The immune system eliminates microbial parasites largely by recognizing and reacting to the antigens of the parasites.

asymptomatic infection An infection that does not generate symptoms.

atherosclerosis A disease of arteries characterized by fatty deposits in the arterial walls. When the deposits become large they may rupture the arterial lining, releasing the fatty material, which may then travel in the blood and block small arteries. By depriving tissue of oxygen, this blockage may cause heart attacks, which result when oxygen-starved heart muscles stop working, or strokes, which result when oxygen-starved brain cells stop working.

autoimmunity The process by which the immune system destroys the body it is supposed to protect. When researchers find evidence of autoimmunity in a disease, they often conclude that the immune system just self-destructed because of its complexity. An electrical engineer involved in reverse engineering might similarly conclude that self-destruct circuits are just an inherent problem that occurs because sophisticated devices malfunction. A role for infection has been generally accepted for several autoimmune diseases, such as rheumatic fever and some kinds of arthritis; rejection of such a role is not warranted by the evidence for any common autoimmune disease.

AZT (azidodideoxythymidine, or zidovudine) A molecule formed from two molecules of a modified building block of nucleic acids (thymine). AZT inhibits HIV by interfering with HIV's reverse transcriptase.

benignity A state of parasitism in which hosts are harmed little. The mistaken assumption that all infectious diseases inexorably evolve toward benignity may prove to be one of the most costly errors in the philosophy of medicine because this assumption guided researchers away from recognizing that evolution toward benignity can be facilitated by intervening in ways that remove the competitive advantages pathogens gain through harmful disease processes.

CDC, or Centers for Disease Control and Prevention The U.S. governmental agency responsible for controlling disease. Implicit in this name is the importance of understanding what a disease is, and what can or cannot be controlled. The increased recognition that chronic conditions may be caused by infections will greatly expand the scope of CDC activities because recognition of infectious causation solidifies categorization of a condition as a disease and reveals new possibilities for control of chronic diseases through control of their infectious agents.

cell culture Cells growing on artificial nutrient media under laboratory conditions. Researchers infect cell cultures in order to study pathogens and their interactions with host cells.

cervix The buttonlike barrier between the vagina and the uterus. Sperm cells must swim through the narrow passageway in the center of the cervix to reach the cavernous uterus, and then continue swimming to the entranceway of one of the oviducts on the opposite side. They continue upstream into the tubelike oviduct to reach the egg cell, which is moving downstream toward the uterus.

childbed fever A disease that often afflicted and killed mothers soon after childbirth until its primary cause, streptococcal infection, was discovered. Also called puerperal fever, it was transmitted to expectant mothers particularly on the ungloved and unwashed hands of physicians and other attendants who picked up the causative organism from postmortem examinations or from infected living patients.

chronic infectious diseases Infectious diseases characterized by prolonged duration. Chronic diseases often have progressively debilitating effects.

cultural vector Refers to a set of characteristics that allows transmission from immobilized hosts to susceptibles, when at least one of the characteristics is some aspect of human culture. Attendants moving pathogens into water supplies or through hospital environments are considered to be components of cultural vectors because such attendants, like mosquitoes and other biological vectors, move pathogens from infectious immobilized hosts to susceptible hosts.

cytotoxic T cells Lymphocytes that attack and destroy infected cells by recognizing antigens on their surfaces. The demolition experts of the immune system.

DNA, or deoxyribonucleic acid A long-chain molecule that encodes information by variations in the sequences of the four different building blocks (deoxyribonucleotides) used along its length. In most organisms the DNA sequences comprise the genes, the primary library of information that is necessary for the development and maintenance of the organism.

endemic A pattern of disease spread in which the disease smolders along within a population, as compared with being spread in an outbreak fashion among populations.

endogenous retroviruses Remnants of retroviruses that were once freely transmissible as viruses. Human endogenous retroviruses descended from retroviruses that once infected our distant human or prehuman ancestors much as HIV and HTLV infect humans today. Endogenous retroviruses are now shipwrecked in a cell's genetic material, transmitted from generation to generation in sperm and egg.

epidemic A disease outbreak that spreads sufficiently to affect simultaneously the different local groups in a geographic region. Narrowly defined, *epidemic* refers to outbreaks in human populations and *epizootic* to outbreaks in populations of other animals. In this book I use the term *epidemic* broadly, to encompass diseases of human and nonhuman hosts.

epidemiologist A scientist who attempts to understand, describe, and explain the distributions of disease in space and time; an epidemiologist is to a physician what a criminologist is to a prosecuting attorney.

epidemiology A scientific discipline that investigates the prevalence and spread of diseases within and among populations of hosts. Epidemiology encompasses any aspect of the environment, the host, or the parasite that is relevant to the prevalence or spread of disease. John Snow's statistical analysis of the spread of cholera in London during the middle of the nineteenth century is the classic study of epidemiology.

fitness The measure of the evolutionary success of an organism relative to that of competitors within the same species. Darwin envisioned this success in terms of the improved fit between the organism and its environment that results from the differential success of competing individuals. Modern evolutionary biologists envision this success in terms of the relative frequencies of alternative genetic instructions. Misunderstanding of this concept and the natural selection that results from differences in fitness is responsible for much confusion in the health sciences over evolutionary processes.

gene A stretch of genetic material (RNA or DNA) that encodes information for a specific protein or functions to regulate the reading of such stretches of genetic material. By such functions, the information encoded in genes has effects on the structure and function of entire organisms and on the inheritance of that structure and function from generation to generation.

genotype The particular set of genes within the organism.

group selection The favoring of a characteristic as a consequence of the differential success of groups of organisms. Group selection is typically weak relative to the differences between the success of individuals within the group, but if groups are small and mix regularly, and if the individuals in groups are genetically related, group selection may be important. These conditions seem especially applicable to groups of parasites inside individual hosts.

helper T cells A class of lymphocytes that influence the activity of other white blood cells and are characterized by the presence of CD4 molecules as receptors protruding from their outer membranes.

HIV Abbreviation for the human immunodeficiency virus.

HTLV Abbreviation for a retrovirus that is commonly known by several names—human T-cell lymphotropic virus, human T-cell leukemia virus, and human T-cell leukemia/lymphoma virus.

immune system The cells of the body that protect against infection by means of generalized responses such as inflammation and by means of specific responses such as the attachment of antibodies to antigens and the destruction of infected cells through recognition of foreign antigens presented on the cells' surfaces.

incidence The number of people that become infected over a given period of time relative to the total size of the population.

inclusive fitness The effect of a characteristic on the passing on of genetic instructions for the characteristic. The term was introduced to emphasize that the evolutionary success of genetic instructions must include the passing on of the instructions through relatives as well as through an individual's own offspring. Evolutionary biologists often use the term *fitness* broadly to encompass inclusive fitness, even though a narrow definition of fitness refers only to the success achieved through an individual's own reproductive activity.

infection The invasion of an organism by a pathogen that completes at least part of its life cycle in the host organism.

infectious disease A disease caused by a pathogen.

inflammatory response A reaction of the lymphatic system and local tissues elicited by parasites or other irritants and characterized by reddening and swelling as blood collects in dilated vessels and lymphatic fluids accumulate in the tissues. Pathogen-destroying white blood cells move from the blood into the tissues during the response.

latency A quiescent state of infection in which a pathogen does not reproduce or reproduces very sluggishly.

lymphocytes A heterogeneous category of white blood cells with specialized immunological functions such as antibody production, destruction of infected cells, and stimulation of other white blood cells.

macrophages White blood cells that engulf parasites, digest them, and present bits of them to other cells to stimulate a broad array of specific immunological responses.

malaria A disease caused by a protozoan in the genus *Plasmodium*, which infects and destroys red blood cells.

mutation A change in the sequence of building blocks (nucleotides) that make up the genes.

mutualism A symbiotic relationship that benefits the species involved.

natural selection The process of differential survival and reproduction that results in changes in gene frequencies and in the characteristics that the genes encode.

nucleic acid Chainlike molecules that are composed of nucleotides and encode genetic information.

nucleotide Building blocks of nucleic acids. Each nucleotide consists of a sugar attached to a phosphoric acid and one of four different nitrogen-containing ringed molecules. In DNA these molecules are adenine, thymine, cytosine, and guanine. In RNA uracil is used in place of thymine, the sugar is a ribose (instead of a deoxyribose), and the nucleotide is called a ribonucleotide. In both RNA and DNA, the sequences of the four molecules encode information.

oncogenes Genes that promote tumor development. Future medical historians may single out the interpretations of the oncogene data during the twentieth century as having grossly misguided our understanding of cancer by causing inappropriate rejection of infectious causation. The presence of oncogenes does not negate the hypothesis of infectious causation of cancers because infectious causes of cancer regularly involve oncogenes as an important part of their cancer-causing mechanisms. The misguided interpretations are akin to saying that a car's open windows are responsible for lost money without taking into account the role of thieves. It could be that money was lost because of open windows, if, for example, wind blew the money out the window. But dismissing the possible role of thieves is unwise unless the arguments invoking thievery can be rejected. If the money was lost while the car was speeding down the highway with open windows and no passengers, then the thievery hypothesis could reliably be rejected.

oviduct The tube that extends from an ovary to the uterus. Egg cells coast downstream in the oviduct toward the uterus, and sperm cells swim upstream in the oviduct toward the ovaries. The twain typically meet in the oviduct.

pandemic A disease outbreak that spreads continentally or worldwide.

papillomaviruses A group of viruses that have DNA as their genetic material. Some types of papillomaviruses cause warts, and others cause cervical cancer and probably several other kinds of cancer.

parasite An individual that lives in or on another individual, and lowers the host individual's fitness. I use "parasite" broadly to encompass pathogens as well as multicellular parasites.

parasitism A symbiotic association in which one member benefits at the expense of the other.

pathogen Something that causes disease. In this book I restrict the use of *pathogen* to a living agent that is at or below the single-cell level of organization. Most pathogens are bacteria, viruses, protozoa, or fungi.

pathogenesis The process by which disease organisms generate disease.

plague A very harmful and widespread disease. Those plagues that are now well understood all involve infectious agents. The various lines of evidence presented in this book indicate that the plagues we do not yet understand—the plagues of cancer, heart disease, mental illness, neurological disease, immunological disease, and infertility—will also turn out to be caused largely by infection.

plasmid A loop of DNA that can occur in bacteria and can be transferred between bacteria. Plasmids can contain genes that code for toxin production, antibiotic resistance, and abilities to invade host cells.

plasmodium The protozoal agent of malaria.

potential for sexual transmission The potential for the spread of sexually transmitted pathogens within a population; this potential subsumes the rate at which sexual contact with new partners occurs and the opportunities for transmission per sexual contact. Use of condoms, for example, reduces the potential for sexual transmission, as does abstinence and increased partner fidelity.

prevalence The extent to which a host population is infected with a parasite.

protein A molecule composed of long chains of amino acids. The associations of different parts of protein chains give proteins a three-dimensional structure that allows them to act as building blocks for organisms and regulate biological processes.

proximate Refers to biological explanations that deal with the mechanics of life processes: how living things function. (Compare "ultimate.")

pulmonary Pertaining to the lungs.

retrovirus A member of a family of viruses that uses RNA as genetic information and transcribes this information into DNA using reverse transcriptase. The family Retroviridae includes HIV, which is in the subfamily Lentivirinae, and HTLV, which is in the subfamily Oncovirinae.

reverse transcriptase An enzyme found in HIV and other viruses that use RNA to encode their genetic information. Reverse transcriptase uses the virus's RNA as a template to form a complementary strand of DNA and then replaces the original RNA strand with a DNA strand that is complementary to the first DNA strand.

risk factor A term for expressing an association between the occurrence of a variable and the occurrence of disease. The term is attractive to some because it does not tread on the dangerous ice of cause and effect. Another advantage is that it serves as a basis for quantifying the relative risks associated with different variables. A disadvantage is that it is useful primarily insofar as it implicates causal factors—something that can be changed to reduce the risk of disease; however, because researchers who use it often tend to stay on the thick ice (describing risk factors instead of interpreting them in a causal framework), rigid risk factor thinking can impede progress toward a detailed understanding of disease causation and, hence, disease control.

RNA, or ribonucleic acid A long-chain molecule that encodes information by variations in the sequences of the four different building blocks (ribonucleotides) used along its length. In some viruses (like human immunodeficiency viruses, hepatitis viruses, and influenza viruses), RNA and the information it encodes comprise the genes—the library of viral information that is necessary for viral invasion and replication. For most organisms RNA serves as a messenger that carries the information encoded by DNA to the sites of protein synthesis, where it then interacts with other kinds of RNA to link amino acids to form proteins.

sepsis The presence of bacteria in the blood or body fluids. Before the identification of bacteria, sepsis referred to the putrefaction of body tissue. The word *antiseptic* is derived from the use of compounds to control the putrefaction of tissues, such as Joseph Lister's use of carbolic acid during the 1860s. A modern remnant of this legacy is the antiseptic mouthwash Listerine.

seroconversion The process by which a host develops antibodies in response to an antigen. The conversion from a negligible to a distinct amount of antibody is usually assessed in the blood serum, hence the name seroconversion.

seropositive The state of having a distinct antibody response that is detectable in the serum.

serum The fluid portion of the blood.

simian A monkey or nonhuman ape.

SIV Abbreviation for simian immunodeficiency virus. SIVs are retroviruses isolated from simian hosts.

strain Formally, an isolate of pathogen that is propagated in the laboratory. I use the term somewhat more loosely to refer to pathogens that are recently derived from a common lineage and are very similar or identical genetically.

symptom A perceptible change in the host that indicates disease. In medical circles, *symptoms* are often defined as subjective manifestations of disease, whereas *signs* are defined as objective manifestations; in this book I use *symptoms* more broadly to include both objective and subjective manifestations.

T cell A category of white blood cells that have diversified to perform a variety of functions, such as stimulating antibody production from other cells and destroying infected cells. Their "T" designation comes from the thymus, which is the organ in which the cells mature.

toxoplasma A protozoan related to the malaria parasite that uses cats as its definitive host, much as the malaria organism uses mosquitoes. It is benign in cats but can cause miscarriages and brain infections in people. The brain infections are responsible for dementia in AIDS patients and may be a primary cause of schizophrenia.

transmission The process by which a parasite moves from one host to another.

ultimate Refers to biological explanations that deal with evolutionary origins: why organisms have the characteristics they have. (Compare "proximate.")

vaccine A preparation that includes antigens of a pathogen and is administered to stimulate the immune system to respond quickly to a pathogen when the pathogen next invades.

vector Something that transports a disease organism from one host to another. Biological vectors include mosquitoes, tsetse flies, sand flies, kissing bugs, fleas, ticks, and lice.

vertebrates Animals with backbones: mammals, birds, reptiles, amphibians, and fishes.

virulence The magnitude of the negative effect of a parasite on its host.

virulence antigen strategy A vaccination strategy that preferentially uses virulence antigens—that is, antigens that make mild organisms more harmful. The goal of the virulence antigen strategy is to make the target pathogen evolve toward benignity and thereby achieve a more complete control of disease than would be possible if the goal were simply to come as close as possible to eradication of the organism.

virus A parasitic organism composed of a protein coat, nucleic acids enclosed by the coat, and sometimes a few enclosed proteins that facilitate infection of host cells. Attention of biologists was distracted for nearly a century by arguments over whether viruses are organisms. The disagreement stems largely from the generalization put forth in the latter half of the nineteenth century that cells are the building blocks of all life. Viruses are simpler than cells, so, the logic goes, viruses cannot be living organisms. This viewpoint seems best dismissed as semantic dog wagging by the tails of dogma.

white blood cells, or leukocytes The cells of the immune system, which circulate in blood, lymph, and body tissues and directly or indirectly destroy invading organisms and infected or damaged cells.

SUGGESTIONS FOR FURTHER READING

Alibek, K. *Biohazard.* New York: Random House, 1999.

Allegra, L., and F. Blasi. Chlamydia pneumoniae *Infection.* Berlin: Springer, 1995.

Burnet, F. M., and D. O. White. *Natural History of Infectious Disease.* Cambridge: Cambridge University Press, 1972.

Diamond, J. *Guns, Germs and Steel.* New York: W. W. Norton, 1997.

Essex, M., et al. *AIDS in Africa.* New York: Raven Press, 1994.

Ewald, P. W. *Evolution of Infectious Disease.* New York: Oxford University Press, 1994.

Garrett, L. *The Coming Plague.* New York: Farrar, Straus and Giroux, 1994.

Hooper, E. *The River: A Journey to the Source of HIV and AIDS.* Boston: Little, Brown & Co., 1999.

Jamison, K. R. *Touched with Fire.* New York: The Free Press, 1993.

McNeill, W. H. *Plagues and Peoples.* Garden City, N.J.: Anchor, 1976.

Morse, S. *Emerging Viruses.* New York: Oxford University Press, 1993.

Nesse, R. M., and G. C. Williams. *Why We Get Sick: The New Science of Darwinian Medicine.* New York: Times Books, 1994.

Northfield, T. C., M. Mendall, and P. M. Goggin. Helicobacter pylori *Infection: Pathophysiology, Epidemiology, and Management.* Deventer, Netherlands: Kluwer, 1993.

Parsonnet, J. *Microbes and Malignancy: Infections As a Cause of Human Cancer.* New York: Oxford University Press, 1999.

Smith, T. *Parasitism and Disease.* Princeton, N.J.: Princeton University Press, 1934.

Stearns, S. C. *Evolution in Health & Disease.* Oxford: Oxford University Press, 1999.

Trevathan, W. R., et al. *Evolutionary Medicine.* New York: Oxford University Press, 1999.

ACKNOWLEDGMENTS

The Nuts and Bolts

This book has benefited immensely from Gregory M. Cochran's generous and enthusiastic sharing of ideas and information. His insights into the causation of chronic diseases helped direct the logic of the book in a fundamental way toward the conclusion that infection is at the root of the major chronic diseases of our time. Without his creative and ingenious input, the chapters on chronic diseases (particularly chapters 3, 8, and 10) would not have been. His encyclopedic mind also contributed critical details to the chapters on biological weapons and the prepared mind (chapters 12 and 15). Greg also served as my de facto expert on science fiction when I wanted to know who said what, when, and where (Chapter 15).

For helpful discussions and access to unpublished material, I thank Robert Genco, Thomas Grayston, Alan Hudson, Philip Ross, and Fuller Torrey. Many of my students passed along relevant references, contributed enthusiastically to discussions, and perhaps even humored me with attentiveness when I droned on about some new idea—thanks in this regard go to Wahid Chammas, Alissa and Jill Saunders, Jeremy Sussman, Dan Rubin, Bruno Walther, Amanda Lang, Edwin Macharia, and Mary Whittle. Melinda LeLacheur of Amherst History Museum graciously provided details about the town of Amherst during the early nineteenth century. Anne Nolan's expert and insightful copyediting improved both accuracy and presentation, as did comments on the penultimate draft by Lynne Gellatly White. The project was supported generously by the Leonard X. Bosack and Bette M. Kruger Charitable Foundation and by Amherst College through its Faculty Research Award Program and its awarding of a fellowship for sabbatical leave.

The final shape of the book owes much to the vision of Free Press editor Stephen Morrow, who tactfully and enthusiastically provided suggestions and kept me from letting the book fall irrecoverably behind schedule. I am glad for his sake that infection rather than stress is the primary cause of peptic ulcers.

A Farewell to W. D. Hamilton

In a touching and vivid obituary, Richard Dawkins mentioned how one of science's greatest thinkers was frequently knocked off his bicycle, probably because Oxford motorists "couldn't believe a man of his age with a great shock of white hair could possibly cycle so fast." Bill Hamilton's shock was already gray during the few years when I knew him on a day-to-day basis. He was in his mid-40s and had just come to the University of Michigan. He came in part because he felt that people in the United States were more enthusiastic about his work than were his peers and superiors in the United Kingdom. There were a few exceptions—people such as Richard Dawkins—but by and large Bill felt that he was living in self-imposed exile. The exile was short-lived, as it turned out—after spending several years at Ann Arbor, he was offered a Royal Society professorship at Oxford, where he could let his mind wander freely, constrained only by a few unavoidable realities, such as car traffic.

I cannot adequately put into words my appreciation for Bill's positive influence on this book, partly because it was his indirect influence on the worker more than his direct influence on the work. I could mention several specific instances that have generated my sense of indebtedness to him—he had a knack for giving soft-spoken yet powerful words of encouragement and knowing when to give them. But for me

Bill's distinctiveness was captured best in an informal presentation that he gave to students at his house in Ann Arbor on a warm autumn evening in 1982.

About fifteen minutes into his presentation he turned off the slide projector and moved over to the blackboard; it was one of the portable, rummage sale kind, with a black slate the size of a place mat mounted on a rickety wooden easel. He looked at the long wooden tray below the board, on the carpet below the tray, and then around the room like someone looking for a lost set of keys. After surveying the situation for about fifteen seconds, he mentioned to us that the children must have taken the chalk. He walked to the back door, stepped outside, holding the door three-quarters open, and called, "Ro-wie . . . Ro-wie . . ."

It was not a voice of irritation. Rather, it was like the voice of a parent calling a child in for dinner, with just enough volume to cover the distance.

I think I heard the wisp of a child's voice coming back on the fall breeze. I pictured her answering but not looking up from her more pressing business, in a sandbox perhaps.

Another drawn-out call came from Bill, with each word carefully but benevolently enunciated: "Did you take the chalk from the blackboard?"

Another pause, and again a child's voice returning from calling distance but in a conversational volume.

Another call with a gentle patience born of perspective rather than training: "I need to use the blackboard for something I want to show to my visitors. I can't find the chalk. Did you do something with the chalk?"

After another barely audible murmur returned with the breeze, Bill walked out of the house to improve the communication. About three minutes later he came back through the living room and smiled at his guests, saying something like, "I think Rowie left the chalk in the playroom."

He disappeared behind us. After a few more minutes, he returned with a nugget of gold in his hands and began drawing some graphs on the chalkboard.

This scene is one of many that flooded back from memory when I learned that Bill had died. He was talking with his daughter on this occasion, but it could have been with a postdoc, a grad student, an admiring colleague, someone who was being unfairly attacked by others, or even a critic who was unfairly attacking him. He treated all with generosity. I never saw him pulling rank. What was so endearing—and what will be sorely missed—is the higher standard that he brought to scientific inquiry, not just the standard of the scientific work but the sense of respect for the people with whom and by whom the enterprise is carried out.

In retrospect it is clear that Bill was always doing or thinking important things, things that would fundamentally change the way we look at nature, and ourselves. Yet he was so generous with his time and so unassuming that those around him could easily forget that this man was in the midst of solving many of nature's great puzzles. When the diversion was over he would pick up where he left off, just as he began writing at the rickety blackboard, as though he had never been distracted by lost chalk.

This book is dedicated with gratitude and admiration to William D. Hamilton. Who would have thought he could ride so fast?

INDEX

Index

Index

Index

Index

Index